The Best American Science Writing 2009

The Best American Science Writing

Editors

2000: James Gleick
2001: Timothy Ferris
2002: Matt Ridley
2003: Oliver Sacks
2004: Dava Sobel
2005: Alan Lightman
2006: Atul Gawande
2007: Gina Kolata
2008: Sylvia Nasar

The Best American Science Writing 2009

EDITOR: NATALIE ANGIER

Series Editor: Jesse Cohen

HARPER PERENNIAL

NEW YORK • LONDON • TORONTO • SYDNEY • NEW DELHI • AUCKLAND

Permissions appear on pages 345–46.

HarperCollins books may be purchased for educational, business, or sales promotional use. For information, please write: Special Markets Department, HarperCollins Publishers, 10 East 53rd Street, New York, NY 10022.

FIRST EDITION

Library of Congress Cataloging-in-Publication Data is available upon request.

ISBN: 978-0-06-143166-1

10 11 12 13 OV/RRD 10 9 8 7 6 5 4 3

Contents

Introduction by Natalie Angier

One of the commonest queries readers have asked me through the years is a variant on the following: "Dear Ms. Angier, I am just finishing up my studies in _____ field of research, and I've decided that what I really want to do is write about science for a popular audience. Any advice on how I should proceed?"

These days, I don't know where to begin, or whom to scold more vigorously: the supplicant or myself. Dear Reader, you want to be a science writer? Have you lost your carbon-based buckeyballs? This is a dreadful time to think about entering the industry! How will you earn a living? Who will hire you? Newspaper science coverage is contracting at an even faster clip than are newspapers overall. Twenty years ago, close to 100 American newspapers proudly displayed their dedicated science and health pages; today, only a couple of dozen science sections remain, and whoops, there goes the one in the *Dallas Morning News*, the *Baltimore Sun*, the *Boston Globe*. Magazines are starved for ads, books are starved for readers, and though people supposedly still watch hours of television a day, the broadcast industry has also been brutal with its clippers. On one particularly bleak day in December of 2008, CNN fired its entire unit of science, environment and technology correspondents; who says a major news network needs anybody on the premises who might know the difference between a blastocyst and a bassoonist? According to a survey of nearly 500 science journalists that was published in March in the journal *Nature*, more than a third of the American respondents were working for operations that had recently cut back on staff, and another third for employers who weren't doing any hiring—and those statistics had been compiled before the latest global economic meltdown. True, the number of science-themed websites and blogs is on the rise, but it remains nearly impossible to make a living off the Internet. All of which suggests that if you plan on being a science writer, you'd better have a few other skills to fall back on—like photosynthesis.

Yet as fun as it is to monger doom, I soon turn the hectoring inward. Aren't you being a bit selfish, Natalie? Can you blame anyone for wanting to be a science writer, for yearning to take a crack at what is arguably the world's greatest profession, short of becoming Thor or Athena? Writing may be terrible and lonely and twisted, and Thomas Mann was right when he said that a writer is a person for whom writing is more difficult than it is for other people, and Red Smith had a point when he compared writing to opening a vein, but as long as you're going to be a writer, you can't do better than to write about science. If you write about science, you will never be bored. You will always be presented with new brain toys, new insights into how the universe is put together. Every time I feel gloomy—about politics, the economy, the wars in Iraq and Afghanistan, the stiff tag on my sweater that is poking into the back of my neck like a plastic sandwich sword and why do clothing manufacturers do that, anyway?—I will talk to a scientist or read a report in a journal, and I will learn something fabulous that makes me grateful for sentience. I'll learn, for example, that the image of your face on a mirror is only half the size of your real face, and that it will stay half the size of your face no matter how far from or near to the mirror you may be. Or I will learn about an extraordinary case of mimicry recently discovered in the Arizona desert, in which hundreds of newborn blister beetle larvae will band together like dancers in a Busby Berkeley musical, jointly assuming the fetching oval shape and striped pattern of a female solitary bee, and moving in lockstep up and down blades of grass, just like a female bee. Their goal? To attract a male bee that will try to mate with the communally fabricated imposter and will thereby unwittingly deliver the larval mass to a bee's nest, where the blister babies can decamp and gorge themselves on nectar, pollen and bee eggs. Or I will learn that the most vocally inventive animal on earth, apart from our ever babbling selves, may well be the walrus, and that the best way to make friends with a walrus is to take a deep breath and then blow in its face.

Aware as I am of the deep pleasures that come with a career in science writing, I can't very well stand scowling at the gate and warn others away. After all, whatever the slumps and surges of the economy, whatever the upheavals and subductions in the media, science marches forward. Science is the future, science is making the future, and nations large and small are busy making future scientists. The more science that emerges from this transplanetary investment in the care and feeding of scientists, the greater the need for the rest of us to sit up and pay attention, to follow the gist of the science with sufficient understanding that we can all have a say in how its fruits will be used. In other words, if we the laity are to keep pace with our labs, we need more science writers, not fewer, and we need more science writing that is clear, wise and eloquent, and that demands to be read. People too often feel estranged from science, convinced that it takes an advanced degree not only to be a scientist, but to understand what scientists do. As a result, they defensively shrug off the whole business as a rarefied geek boutique of little relevance to their lives. I am convinced that one of the surest cures for scientific illiteracy is great scientific literature, writing that doesn't merely translate technical jargon into English or explain arcane ideas simply—the usual tasks we associate with science writing—but that transforms its subject from peripheral to central, from alien abstraction to personal revelation. A great piece of science writing goes beyond illuminating the topic it is nominally about. It throws wide the door to the entire scientific enterprise and invites you inside. It gives you a sense of entitlement. You've met science, you've understood it, you're ready to do it again. Science becomes like music or art. Sure, you may not be able to play an instrument or draw a straight line. But you can trust your ears and eyes, and your hungry, inquiring mind.

So when I finally move from dire hand-wringings to practical advice, here is what I tell aspiring science writers:

> If you haven't done it already you should earn a science Ph.D., not because you need the degree to understand the material but because I've been amazed, throughout my 30-year career, to observe how much a doctorate can enhance one's professional credibility at every stage of life, and it's better to get it over with when you're young enough to tolerate being an overtaxed, underpaid, and marginally respected slave labeler of petri dishes and scrubber of flasks, which I'm sorry to say is what graduate students are.
>
> When you are trying to describe difficult things you should appeal to as many senses as possible, to ask, what does it look like, sound like, feel like, smell like. One of the most evocative responses I got from a scientist was Cynthia Wolberger's reply to my question, what would a body cell look like if blown up to the size of a desktop accessory? Cells are very gooey and viscous, she said gaily, "so it would probably look like snot." So now you know.
>
> My brother Joe once gave me an all-purpose piece of advice that I have trouble adhering to but still think it brilliant enough to pass on profligately: Do not believe either extreme criticism or exalted praise, or you will soon be paralyzed in the face of one and unable to function without the other.
>
> There's no way around it. Good writing takes time. Don't

beat yourself up about it, but be sure to figure it into your schedule.

Writers must read, read, read, and if you're an aspiring science writer or an aspiring science lover, or simply a lover of great stories beautifully told, I can't imagine a better place to turn than to this collection.

The articles gathered here exemplify science writing at full, thrilling gallop, and though it might be as much of a transgression against Joe's Law to express wild praise as it is to crave it, I must confess that I deeply admire all these stories and love quite a few. They cover a wide range of disciplines—medicine, the environment, evolutionary biology, computer science, chemistry, physics, astronomy—and they address topics both timely and timeless, specific and universal. Atul Gawande prompts us to consider the familiar yet oddly elusive sensation that is itchiness. It is not really painful, and in fact it can be pleasurable to scratch an itch, but after a certain point the scratching makes the itchiness spread further and go deeper, and you keep scratching and scratching, desperate to claw the itch out, until finally you're on the verge of baying at the moon. The story includes one of the most gruesome moments in the annals of medical writing, when the principal character, a fortyish woman known as M., wakes up with greenish fluid dripping down her face. Its source? "She had scratched through her skull during the night," Gawande writes, "and all the way into her brain."

Just as the circuitry of itchiness can be tricky to trace, and psychosocial elements clearly play a role—studies have shown that the scratch reflex is almost as publicly contagious as yawning—so too is the body's pain response a complex weave of the physical and the perceptual, of pain fibers stimulated and pain signals codified. Annie Murphy Paul asks the difficult and potentially incendiary question of when a growing baby first begins to feel pain. In the womb? If so, at what stage of gestation? And should evidence of fetal pain affect when and how abortions are performed? Jina Moore considers how best to tally unimaginable pain, describing physicians who treat victims of torture, measuring and classifying scars, burns, broken bones badly healed, the better to use the results in international courts of law.

Life brings pain, but must we inflict pain on one another? Are humans intractably savage, doomed to wage war, commit murder, bathe the latest hated Otherland in blood? Several articles take different approaches to ask the same basic question. Primatologist Marina Cords describes the revealing patterns of hostility among neighboring groups of blue monkeys in western Kenya, in which the females, rather than the males, do the bulk of the fighting, and oh they can be brutal. After one encounter, Dr. Cords noticed that a female's leg had been ripped open, and "the skin bunched down at her ankle like a frilly bobby sock, leaving the red calf muscle fully exposed." Oddly enough, although the outcome of a territorial dispute will affect all the monkeys in the warring groups, only a handful of the animals serve as soldiers from each clan, and Dr. Cords explains who chooses to fight and why they bother. John Horgan talks with many of the bigger names and egos in evolutionary biology and anthropology—Frans de Waal, Richard Wrangham, Robert Sapolsky, Edward O. Wilson—and concludes that, if we hominids could somehow manage to equitably distribute food, fuel, and other essential resources and elevate the position of women worldwide, war might well be, as Horgan felicitously puts it, evitable.

If eliminating wars worldwide sounds overly ambitious, we might find novel ways to stanch the blood on a smaller scale. In "Blocking the Transmission of Violence," Alex Kotlowitz describes the half-heroic, half-crankily quixotic efforts of the epidemiologist Gary Slutkin to reduce the inner-city murder rate by treating the problem like an infectious disease. Forget about overarching socioeconomic themes, Slutkin argues, forget about endemic poverty, lack of job opportunities, decaying schools, racism. If you want to keep people from murdering each other, you have to identify the nodes of infection, and step in point by point, and aggressively quarantine.

Another condition that we don't often consider infectious is cancer, yet as David Quammen masterfully elucidates, there are contagious tumors, and there are sound Darwinian reasons why such tumors exist. He tells the story of Devil Facial Tumor Disease, a cancer that is devastating Tasmanian devils, the fierce, piglet-sized marsupials found only on Tasmania. The devils transmit the tumors to one another through biting and other rough interactions, and the tumors, "crumbly, like feta cheese," grow quickly, "filling eye sockets, distending cheeks, making it difficult for the animals to see or to eat." Quammen also talks of other cases in which cancer proved transmissible, like the time a laboratory worker accidentally jabbed herself with a syringe of colon cancer cells. A colonic tumor grew in her hand, but unlike the poor devils, she could be rescued by surgery.

Quammen then lays out the rise of multicellularity. "Cooperation was a winning formula," he writes. "Primitive multicellular creatures, roughly along the lines of jellyfish or sponges or slime molds, began to succeed, to grow, to occupy space, and to claim resources in ways that loner cells couldn't." Yet still our bodies are ever at risk that cells will turn rogue, will reclaim their primal roots and begin dividing lawlessly. Cancerous cells normally are limited in their ambitions; when their host dies of their malignant behaviors, they die, too. Contagious cancers overleap this boundary. The surprise is that we haven't seen more of them.

Irene Pepperberg is striving to overleap a very different sort of barrier, the one dividing us from other species. Through her celebrated work with Alex, the African gray parrot, as well as with his less obviously gifted replacements, Pepperberg has explored the outermost precincts of animal language and sought a glimpse of life reframed from a bird's-eye view. Margaret Talbot conveys Pepperberg's odyssey with perfect pitch. "In speaking about animal intelligence, Pepperberg has tried to strike a balance between what the ecologist James Gould has called 'the unprofitable extremes of blinding skepticism and crippling romanticism,'" Talbot writes. "Soon enough, 'birdbrain' may no longer be a viable insult."

Nor are we content with our human brains, which is why we are desperate to improve them, through algorithms like that promoted by Piotr Wozniak, "inventor of a technique to turn people into geniuses," writes Gary Wolf, as long as we're willing to turn our lives over to a rigidly floral sort of iCal; and why we want to humanize our computers, grant them the power to follow our every rambling utterance and make sense of the most garbled twang: John Seabrook plaintively asks, "Will we ever get a computer we can really talk to?" Then again, is there any way to get our computers to shut up? In "The Anonymity Experiment," Catherine Price lays out her week-long effort "to hide in plain sight," to reclaim her privacy, be "as anonymous as possible while still living a normal life." Living anonymously meant not using her regular cellphone, landline phone, e-mail account, any of her photo IDs, E-ZPass or supermarket loyalty card, and staying away from government buildings, airports, banks, anywhere with surveillance cameras, and, well, you get the picture, smile! you're on YouTube, and we all get it, too.

Lest you despair for our future as earthlings, with our hacking away at the giraffe's rightful habitat or our need to spin human sewage into potable water—and it's not bad, that recycled effluvia; Elizabeth Royte gulps it down and then, like Oliver Twist, sticks out her cup and asks for more—you can look out to space, where, according to a scientist Karen Olsson quotes, "we're just a bit of pollution," and most of the universe "is made of something else." Dark matter, dark energy, the darkness of death. In her gorgeous essay, "Perhaps Death Is Proud," Theresa Brown tells about her experience as a "new nurse" confronting her first "Condition A," a cardiac arrest. The patient was a "particularly nice older woman with lung cancer," and one minute she was "fine," in that low-grade hospital way, and the next she had blood spilling from her mouth and nose, and she tried to stand up, and she sat right back down, and started shaking and collapsed back onto the bed. Code A! They tried to put a tube down her throat to give her oxygen, but blood blocked the view. They tried cutting a breathing hole right into her trachea, but that filled with blood, too. Her heartbeat line "was wobbly and unformed. . . . She could not be saved." What to do, what lesson to extract? "Go home, love your children, try not to bicker," Brown writes. "Laugh loud and often, as much as possible." It's very simple: "The antidote to death is life."

Yes, laughter is good medicine, and it's fine science, too. For that reason I chose to close the collection with a piece from the *Onion*, which is rightfully famed for its riffs on politics, the economy, dorm-room beer posters, Martin Luther "I Had a Really Weird Dream" King, Jr., frottage, the Area Man and "Emily Saunders, Food Technologist," but is less well known for its surgically virtuosic science satire, like the article in which "Scientists Ask Congress to Fund $50 Billion Science Thing," and another about how "World's Scientists Admit They Just Don't Like Mice," and how about "Dolphins Evolve Opposable Thumbs: 'Oh, Shit,' Says Humanity." If you can laugh loud and often at science, you must really feel at home.

The Best American Science Writing 2009

ATUL GAWANDE

The Itch

FROM *THE NEW YORKER*

What explains pathological itching—the excruciating sensation that never seems to go away, no matter how long you scratch? As Atul Gawande reports, coming up with a useful answer requires a new model of how the brain works.

IT WAS STILL SHOCKING TO M. HOW MUCH A FEW WRONG turns could change your life. She had graduated from Boston College with a degree in psychology, married at twenty-five, and had two children, a son and a daughter. She and her family settled in a town on Massachusetts' southern shore. She worked for thirteen years in health care, becoming the director of a residence program for men who'd suffered severe head injuries. But she and her husband began fighting. There were betrayals. By the time she was thirty-two, her marriage had disintegrated. In the divorce, she lost possession of their home, and, amid her financial and psychological struggles, she saw that she was losing her children, too. Within a few years, she was drinking. She began dating someone, and they drank together. After a while, he brought some drugs home, and she tried them. The drugs got harder. Eventually, they were doing heroin, which turned out to be readily available from a street dealer a block away from her apartment.

One day, she went to see a doctor because she wasn't feeling well, and learned that she had contracted H.I.V. from a contaminated needle. She had to leave her job. She lost visiting rights with her children. And she developed complications from the H.I.V., including shingles, which caused painful, blistering sores across her scalp and forehead. With treatment, though, her H.I.V. was brought under control. At thirty-six, she entered rehab, dropped the boyfriend, and kicked the drugs. She had two good, quiet years in which she began rebuilding her life. Then she got the itch.

It was right after a shingles episode. The blisters and the pain responded, as they usually did, to acyclovir, an antiviral medication. But this time the area of the scalp that was involved became numb, and the pain was replaced by a constant, relentless itch. She felt it mainly on the right side of her head. It crawled along her scalp, and no matter how much she scratched it would not go away. "I felt like my inner self, like my brain itself, was itching," she says. And it took over her life just as she was starting to get it back.

Her internist didn't know what to make of the problem. Itching is an extraordinarily common symptom. All kinds of dermatological conditions can cause it: allergic reactions, bacterial or fungal infections, skin cancer, psoriasis, dandruff, scabies, lice, poison ivy, sun damage, or just dry skin. Creams and makeup can cause itch, too. But M. used ordinary shampoo and soap, no creams. And when the doctor examined M.'s scalp she discovered nothing abnormal—no rash, no redness, no scaling, no thickening, no fungus, no parasites. All she saw was scratch marks.

The internist prescribed a medicated cream, but it didn't help. The urge to scratch was unceasing and irresistible. "I would try to control it during the day, when I was aware of the itch, but it was really hard," M. said. "At night, it was the worst. I guess I would scratch when I was asleep, because in the morning there would be blood on my pillowcase." She began to lose her hair over the itchy area. She returned to her internist again and again. "I just kept haunting her and calling her," M. said. But nothing the internist tried worked, and she began to suspect that the itch had nothing to do with M.'s skin.

Plenty of non-skin conditions can cause itching. Dr. Jeffrey Bernhard, a dermatologist with the University of Massachusetts Medical School, is among the few doctors to study itching systematically (he published the definitive textbook on the subject), and he told me of cases caused by hyperthyroidism, iron deficiency, liver disease, and cancers like Hodgkin's lymphoma. Sometimes the syndrome is very specific. Persistent outer-arm itching that worsens in sunlight is known as *brachioradial pruritus*, and it's caused by a crimped nerve in the neck. *Aquagenic pruritus* is recurrent, intense, diffuse itching upon getting out of a bath or shower, and although no one knows the mechanism, it's a symptom of polycythemia vera, a rare condition in which the body produces too many red blood cells.

But M.'s itch was confined to the right side of her scalp. Her viral count showed that the H.I.V. was quiescent. Additional blood tests and X-rays were normal. So the internist concluded that M.'s problem was probably psychiatric. All sorts of psychiatric conditions can cause itching. Patients with psychosis can have cutaneous delusions—a belief that their skin is infested with, say, parasites, or crawling ants, or laced with tiny bits of fibreglass. Severe stress and other emotional experiences can also give rise to a physical symptom like itching—whether from the body's release of endorphins (natural opioids, which, like morphine, can cause itching), increased skin temperature, nervous scratching, or increased sweating. In M.'s case, the internist suspected tricho-tillomania, an obsessive-compulsive disorder in which patients have an irresistible urge to pull out their hair.

M. was willing to consider such possibilities. Her life had been a mess, after all. But the antidepressant medications often prescribed for O.C.D. made no difference. And she didn't actually feel a compulsion to pull out her hair. She simply felt *itchy*, on the area of her scalp that was left numb from the shingles. Although she could sometimes distract herself from it—by watching television or talking with a friend—the itch did not fluctuate with her mood or level of stress. The only thing that came close to offering relief was to scratch.

"Scratching is one of the sweetest gratifications of nature, and as ready at hand as any," Montaigne wrote. "But repentance follows too annoyingly close at its heels." For M., certainly, it did: the itching was so torturous, and the area so numb, that her scratching began to go through the skin. At a later office visit, her doctor found a silver-dollar-size patch of scalp where skin had been replaced by scab. M. tried bandaging her head, wearing caps to bed. But her fingernails would always find a way to her flesh, especially while she slept.

One morning, after she was awakened by her bedside alarm, she sat up and, she recalled, "this fluid came down my face, this greenish liquid." She pressed a square of gauze to her head and went to see her doctor again. M. showed the doctor the fluid on the dressing. The doctor looked closely at the wound. She shined a light on it and in M.'s eyes. Then she walked out of the room and called an ambulance. Only in the Emergency Department at Massachusetts General Hospital, after the doctors started swarming, and one told her she needed surgery *now*, did M. learn what had happened. She had scratched through her skull during the night—and all the way into her brain.

Itching is a most peculiar and diabolical sensation. The definition offered by the German physician Samuel Hafenreffer in

1660 has yet to be improved upon: An unpleasant sensation that provokes the desire to scratch. Itch has been ranked, by scientific and artistic observers alike, among the most distressing physical sensations one can experience. In Dante's Inferno, falsifiers were punished by "the burning rage / of fierce itching that nothing could relieve":

The way their nails scraped down upon the scabs
Was like a knife scraping off scales from carp. . . .
"O you there tearing at your mail of scabs
And even turning your fingers into pincers,"
My guide began addressing one of them,

"Tell us are there Italians among the souls
Down in this hole and I'll pray that your nails
Will last you in this task eternally."

Though scratching can provide momentary relief, it often makes the itching worse. Dermatologists call this the itch-scratch cycle. Scientists believe that itch, and the accompanying scratch reflex, evolved in order to protect us from insects and clinging plant toxins—from such dangers as malaria, yellow fever, and dengue, transmitted by mosquitoes; from tularemia, river blindness, and sleeping sickness, transmitted by flies; from typhus-bearing lice, plague-bearing fleas, and poisonous spiders. The theory goes a long way toward explaining why itch is so exquisitely tuned. You can spend all day without noticing the feel of your shirt collar on your neck, and yet a single stray thread poking out, or a louse's fine legs brushing by, can set you scratching furiously.

But how, exactly, itch works has been a puzzle. For most of medical history, scientists thought that itching was merely a weak form of pain. Then, in 1987, the German researcher H. O. Handwerker and his colleagues used mild electric pulses to drive histamine, an itch-producing substance that the body releases during allergic reactions, into the skin of volunteers. As the researchers increased the dose of histamine, they found that they were able to increase the intensity of itch the volunteers reported, from the barely appreciable to the "maximum imaginable." Yet the volunteers never felt an increase in pain. The scientists concluded that itch and pain are entirely separate sensations, transmitted along different pathways.

Despite centuries spent mapping the body's nervous circuitry, scientists had never noticed a nerve specific for itch. But now the hunt was on, and a group of Swedish and German researchers embarked upon a series of tricky experiments. They inserted ultra-thin metal electrodes into the skin of paid volunteers, and wiggled them around until they picked up electrical signals from a single nerve fibre. Computers subtracted the noise from other nerve fibres crossing through the region. The researchers would then spend hours—as long as the volunteer could tolerate it—testing different stimuli on the skin in the area (a heated probe, for example, or a fine paintbrush) to see what would get the nerve to fire, and what the person experienced when it did.

They worked their way through fifty-three volunteers. Mostly, they encountered well-known types of nerve fibres that respond to temperature or light touch or mechanical pressure. "That feels warm," a volunteer might say, or "That feels soft," or "Ouch! Hey!" Several times, the scientists came across a nerve fibre that didn't respond to any of these stimuli. When they introduced a tiny dose of histamine into the skin, however, they observed a sharp electrical response in some of these nerve fibres, and the volunteer would experience an itch. They announced their discovery in a 1997 paper: they'd found a type of nerve that was specific for itch.

Unlike, say, the nerve fibres for pain, each of which covers a millimetre-size territory, a single itch fibre can pick up an itchy sensation more than three inches away. The fibres also turned out to have extraordinarily low conduction speeds, which explained why itchiness is so slow to build and so slow to subside.

Other researchers traced these fibres to the spinal cord and all the way to the brain. Examining functional PET-scan studies in healthy human subjects who had been given mosquito-bite-like histamine injections, they found a distinct signature of itch activity. Several specific areas of the brain light up: the part of the cortex that tells you where on your body the sensation occurs; the region that governs your emotional responses, reflecting the disagreeable nature of itch; and the limbic and motor areas that process irresistible urges (such as the urge to use drugs, among the addicted, or to overeat, among the obese), reflecting the ferocious impulse to scratch.

Now various phenomena became clear. Itch, it turns out, is indeed inseparable from the desire to scratch. It can be triggered chemically (by the saliva injected when a mosquito bites, say) or mechanically (from the mosquito's legs, even before it bites). The itch-scratch reflex activates higher levels of your brain than the spinal-cord-level reflex that makes you pull your hand away from a flame. Brain scans also show that scratching diminishes activity in brain areas associated with unpleasant sensations.

But some basic features of itch remained unexplained—features that make itch a uniquely revealing case study. On the one hand, our bodies are studded with receptors for itch, as they are with receptors for touch, pain, and other sensations; this provides an alarm system for harm and allows us to safely navigate the world. But why does a feather brushed across the skin sometimes itch and at other times tickle? (Tickling has a social component: you can make yourself itch, but only another person can tickle you.) And, even more puzzling, how is it that you can make yourself itchy just by thinking about it?

Contemplating what it's like to hold your finger in a flame won't make your finger hurt. But simply writing about a tick crawling up the nape of one's neck is enough to start my neck itching. Then my scalp. And then this one little spot along my flank where I'm beginning to wonder whether I should check to see if there might be something there. In one study, a German professor of psychosomatics gave a lecture that included, in the first half, a series of what might be called itchy slides, showing fleas, lice, people scratching, and the like, and, in the second half, more benign slides, with pictures of soft down, baby skin, bathers. Video cameras recorded the audience. Sure enough, the frequency of scratching among people in the audience increased markedly during the first half and decreased during the second. Thoughts made them itch.

We now have the nerve map for itching, as we do for other sensations. But a deeper puzzle remains: how much of our sensations and experiences do nerves really explain?

In the operating room, a neurosurgeon washed out and debrided M.'s wound, which had become infected. Later, a plastic surgeon covered it with a graft of skin from her thigh. Though her head was wrapped in layers of gauze and she did all she could to resist the still furious itchiness, she awoke one morning to find that she had rubbed the graft away. The doctors returned her to the operating room for a second skin graft, and this time they wrapped her hands as well. She rubbed it away again anyway.

"They kept telling me I had O.C.D.," M. said. A psychiatric team was sent in to see her each day, and the resident would ask her, "As a child, when you walked down the street did you count the lines? Did you do anything repetitive? Did you have to count everything you saw?" She kept telling him no, but he seemed skeptical. He tracked down her family and asked them, but they said no, too. Psychology tests likewise ruled out obsessive-compulsive disorder. They showed depression, though, and, of course, there was the history of addiction. So the doctors still thought her scratching was from a psychiatric disorder. They gave her drugs that made her feel logy and sleep a lot. But the itching was as bad as ever, and she still woke up scratching at that terrible wound.

One morning, she found, as she put it, "this very bright and happy-looking woman standing by my bed. She said, 'I'm Dr. Oaklander,'" M. recalled. "I thought, Oh great. Here we go again. But she explained that she was a neurologist, and she said, 'The first thing I want to say to you is that I don't think you're crazy. I don't think you have O.C.D.' At that moment, I really saw her grow wings and a halo," M. told me. "I said, 'Are you sure?' And she said, 'Yes. I have heard of this before.' "

Anne Louise Oaklander was about the same age as M. Her mother is a prominent neurologist at Albert Einstein College of Medicine, in New York, and she'd followed her into the field. Oaklander had specialized in disorders of peripheral nerve sensation—disorders like shingles. Although pain is the most common symptom of shingles, Oaklander had noticed during her training that some patients also had itching, occasionally severe, and seeing M. reminded her of one of her shingles patients. "I remember standing in a hallway talking to her, and what she complained about—her major concern—was that she was tormented by this terrible itch over the eye where she had had shingles," she told me. When Oaklander looked at her, she thought that something wasn't right. It took a moment to realize why. "The itch was so severe, she had scratched off her eyebrow."

Oaklander tested the skin near M.'s wound. It was numb to temperature, touch, and pinprick. Nonetheless, it was itchy, and when it was scratched or rubbed M. felt the itchiness temporarily subside. Oaklander injected a few drops of local anesthetic into the skin. To M.'s surprise, the itching stopped—instantly and almost entirely. This was the first real relief she'd had in more than a year.

It was an imperfect treatment, though. The itch came back when the anesthetic wore off, and, although Oaklander tried having M. wear an anesthetic patch over the wound, the effect diminished over time. Oaklander did not have an explanation for any of this. When she took a biopsy of the itchy skin, it showed that ninety-six per cent of the nerve fibres were gone. So why was the itch so intense?

Oaklander came up with two theories. The first was that those few remaining nerve fibres were itch fibres and, with no other fibres around to offer competing signals, they had become constantly active. The second theory was the opposite. The nerves were dead, but perhaps the itch system in M.'s brain had gone haywire, running on a loop all its own.

The second theory seemed less likely. If the nerves to her scalp were dead, how would you explain the relief she got from scratching, or from the local anesthetic? Indeed, how could you explain the itch in the first place? An itch without nerve endings didn't make sense. The neurosurgeons stuck with the first theory; they offered to cut the main sensory nerve to the front of M.'s scalp and abolish the itching permanently. Oaklander, however, thought that the second theory was the right one—that this was a brain problem, not a nerve problem—and that cutting the nerve would do more harm than good. She argued with the neurosurgeons, and she advised M. not to let them do any cutting.

"But I was desperate," M. told me. She let them operate on her, slicing the supraorbital nerve above the right eye. When she woke up, a whole section of her forehead was numb—and the itching was gone. A few weeks later, however, it came back, in an even wider expanse than before. The doctors tried pain medications, more psychiatric medications, more local anesthetic. But the only thing that kept M. from tearing her skin and skull open again, the doctors found, was to put a foam football helmet on her head and bind her wrists to the bedrails at night.

She spent the next two years committed to a locked medical ward in a rehabilitation hospital—because, although she was not mentally ill, she was considered a danger to herself. Eventually, the staff worked out a solution that did not require binding her to the bedrails. Along with the football helmet, she had to wear white mitts that were secured around her wrists by surgical tape. "Every bedtime, it looked like they were dressing me up for Halloween—me and the guy next to me," she told me.

"The guy next to you?" I asked. He had had shingles on his neck, she explained, and also developed a persistent itch. "Every night, they would wrap up his hands and wrap up mine." She spoke more softly now. "But I heard he ended up dying from it, because he scratched into his carotid artery."

I met M. seven years after she'd been discharged from the rehabilitation hospital. She is forty-eight now. She lives in a three-room apartment, with a crucifix and a bust of Jesus on the wall and the low yellow light of table lamps strung with beads over their shades. Stacked in a wicker basket next to her coffee table were Rick Warren's "The Purpose Driven Life," *People*, and the latest issue of *Neurology Now*, a magazine for patients. Together, they summed up her struggles, for she is still fighting the meaninglessness, the isolation, and the physiology of her predicament.

She met me at the door in a wheelchair; the injury to her brain had left her partially paralyzed on the left side of her body. She remains estranged from her children. She has not, however, relapsed into drinking or drugs. Her H.I.V. remains under control. Although the itch on her scalp and forehead persists, she has gradually learned to protect herself. She trims her nails short. She finds ways to distract herself. If she must scratch, she tries to rub gently instead. And, if that isn't enough, she uses a soft toothbrush or a rolled-up terry cloth. "I don't use anything sharp," she said. The two years that she spent bound up in the hospital seemed to have broken the nighttime scratching. At home, she found that she didn't need to wear the helmet and gloves anymore.

Still, the itching remains a daily torment. "I don't normally tell people this," she said, "but I have a fantasy of shaving off my eyebrow and taking a metal-wire grill brush and scratching away."

Some of her doctors have not been willing to let go of the idea that this has been a nerve problem all along. A local neurosurgeon told her that the original operation to cut the sensory nerve to her scalp must not have gone deep enough. "He wants to go in again," she told me.

A NEW SCIENTIFIC UNDERSTANDING OF perception has emerged in the past few decades, and it has overturned classical, centuries-long beliefs about how our brains work—though it has apparently not penetrated the medical world yet. The old understanding of perception is what neuroscientists call "the naïve view," and it is the view that most people, in or out of medicine, still have. We're inclined to think that people normally perceive things in the world directly. We believe that the hardness of a rock, the coldness of an ice cube, the itchiness of a sweater are picked up by our nerve endings, transmitted through the spinal cord like a message through a wire, and decoded by the brain.

In a 1710 "Treatise Concerning the Principles of Human Knowledge," the Irish philosopher George Berkeley objected to this view. We do not know the world of objects, he argued; we know only our mental ideas of objects. "Light and colours, heat and cold, extension and figures—in a word, the things we see and feel—what are they but so many sensations, notions, ideas?" Indeed, he concluded, the objects of the world are likely just inventions of the mind, put in there by God. To which Samuel Johnson famously responded by kicking a large stone and declaring, "I refute it *thus*!"

Still, Berkeley had recognized some serious flaws in the direct-perception theory—in the notion that when we see, hear, or feel we are just taking in the sights, sounds, and textures of the world. For one thing, it cannot explain how we experience things that seem physically real but aren't: sensations of itching that arise from nothing more than itchy thoughts; dreams that can seem indistinguishable from reality; phantom sensations that amputees have in their missing limbs. And, the more we examine the actual nerve transmissions we receive from the world outside, the more inadequate they seem.

Our assumption had been that the sensory data we receive from our eyes, ears, nose, fingers, and so on contain all the information that we need for perception, and that perception must work something like a radio. It's hard to conceive that a Boston Symphony Orchestra concert is in a radio wave. But it is. So you might think that it's the same with the signals we receive—that if you hooked up someone's nerves to a monitor you could watch what the person is experiencing as if it were a television show.

Yet, as scientists set about analyzing the signals, they found them to be radically impoverished. Suppose someone is viewing a tree in a clearing. Given simply the transmissions along the optic nerve from the light entering the eye, one would not be able to reconstruct the three-dimensionality, or the distance, or the detail of the bark—attributes that we perceive instantly.

Or consider what neuroscientists call "the binding problem." Tracking a dog as it runs behind a picket fence, all that your eyes receive is separated vertical images of the dog, with large slices missing. Yet somehow you perceive the mutt to be whole, an intact entity travelling through space. Put two dogs together behind the fence and you don't think they've morphed into one. Your mind now configures the slices as two independent creatures.

The images in our mind are extraordinarily rich. We can tell if something is liquid or solid, heavy or light, dead or alive. But the information we work from is poor—a distorted, two-dimensional transmission with entire spots missing. So the mind fills in most of the picture. You can get a sense of this from brain-anatomy studies. If visual sensations were primarily received rather than constructed by the brain, you'd expect that most of the fibres going to the brain's primary visual cortex would come from the retina. Instead, scientists have found that only twenty per cent do; eighty per cent come downward from regions of the brain governing functions like memory. Richard Gregory, a prominent British neuropsychologist, estimates that visual perception is more than ninety per cent memory and less than ten per cent sensory nerve signals. When Oaklander theorized that M.'s itch was endogenous, rather than generated by peripheral nerve signals, she was onto something important.

The fallacy of reducing perception to reception is especially clear when it comes to phantom limbs. Doctors have often explained such sensations as a matter of inflamed or frayed nerve endings in the stump sending aberrant signals to the brain. But this explanation should long ago have been suspect. Efforts by surgeons to cut back on the nerve typically produce the same results that M. had when they cut the sensory nerve to her forehead: a brief period of relief followed by a return of the sensation.

Moreover, the feelings people experience in their phantom limbs are far too varied and rich to be explained by the random firings of a bruised nerve. People report not just pain but also sensations of sweatiness, heat, texture, and movement in a missing limb. There is no experience people have with real limbs that they do not experience with phantom limbs. They feel their phantom leg swinging, water trickling down a phantom arm, a phantom ring becoming too tight for a phantom digit. Children have used phantom fingers to count and solve arithmetic problems. V. S. Ramachandran, an eminent neuroscientist at the University of California, San Diego, has written up the case of a woman who was born with only stumps at her shoulders, and yet, as far back as she could remember, felt herself to have arms and hands; she even feels herself gesticulating when she speaks. And phantoms do not occur just in limbs. Around half of women who have undergone a mastectomy experience a phantom breast, with the nipple being the most vivid part. You've likely had an experience of phantom sensation yourself. When the dentist gives you a local anesthetic, and your lip goes numb, the nerves go dead. Yet you don't feel your lip disappear. Quite the opposite: it feels larger and plumper than normal, even though you can see in a mirror that the size hasn't changed.

The account of perception that's starting to emerge is what we might call the "brain's best guess" theory of perception: perception is the brain's best guess about what is happening in the outside world. The mind integrates scattered, weak, rudimentary signals from a variety of sensory channels, information from past experiences, and hard-wired processes, and produces a sensory experience full of brain-provided color, sound, texture, and meaning. We see a friendly yellow Labrador bounding behind a picket fence not because that is the transmission we receive but because this is the perception our weaver-brain assembles as its best hypothesis of what is out there from the slivers of information we get. Perception is inference.

The theory—and a theory is all it is right now—has begun to make sense of some bewildering phenomena. Among them is an experiment that Ramachandran performed with volunteers who had phantom pain in an amputated arm. They put their surviving arm through a hole in the side of a box with a mirror inside, so that, peering through the open top, they would see their arm and its mirror image, as if they had two arms. Ramachandran then asked them to move both their intact arm and, in their mind, their phantom arm—to pretend that they were conducting an orchestra, say. The patients had the sense that they had two arms again. Even though they knew it was an illusion, it provided immediate relief. People who for years had been unable to unclench their phantom fist suddenly felt their hand open; phantom arms in painfully contorted positions could relax. With daily use of the mirror box over weeks, patients sensed their phantom limbs actually shrink into their stumps and, in several instances, completely vanish. Researchers at Walter Reed Army Medical Center recently published the results of a randomized trial of mirror therapy for soldiers with phantom-limb pain, showing dramatic success.

A lot about this phenomenon remains murky, but here's what the new theory suggests is going on: when your arm is amputated, nerve transmissions are shut off, and the brain's best guess often seems to be that the arm is still there, but paralyzed, or clenched, or beginning to cramp up. Things can stay like this for years. The mirror box, however, provides the brain with new visual input—however illusory—suggesting motion in the absent arm. The brain has to incorporate the new information into its sensory map of what's happening. Therefore, it guesses again, and the pain goes away.

The new theory may also explain what was going on with M.'s itch. The shingles destroyed most of the nerves in her scalp. And, for whatever reason, her brain surmised from what little input it had that something horribly itchy was going on—that perhaps a whole army of ants were crawling back and forth over just that patch of skin. There wasn't any such thing, of course. But M.'s brain has received no contrary signals that would shift its assumptions. So she itches.

NOT LONG AGO, I MET a man who made me wonder whether such phantom sensations are more common than we realize. H. was forty-eight, in good health, an officer at a Boston financial-services company living with his wife in a western suburb, when he made passing mention of an odd pain to his internist. For at least twenty years, he said, he'd had a mild tingling running along his left arm and down the left side of his body, and, if he tilted his neck forward at a particular angle, it became a pronounced, electrical jolt. The internist recognized this as Lhermitte's sign, a classic symptom that can indicate multiple sclerosis, vitamin B_{12} deficiency, or spinal-cord compression from a tumor or a herniated disk. An MRI revealed a cavernous hemangioma, a pea-size mass of dilated blood vessels, pressing into the spinal cord in his neck. A week later, while the doctors were still contemplating what to do, it ruptured.

"I was raking leaves out in the yard and, all of a sudden, there was an explosion of pain and my left arm wasn't responding to my brain," H. said when I visited him at home. Once the swelling subsided, a neurosurgeon performed a tricky operation to remove the tumor from the spinal cord. The operation was successful, but afterward H. began experiencing a constellation of strange sensations. His left hand felt cartoonishly large—at least twice its actual size. He developed a constant burning pain along an inch-wide ribbon extending from the left side of his neck all the way down his arm. And an itch crept up and down along the same band, which no amount of scratching would relieve.

H. has not accepted that these sensations are here to stay—the prospect is too depressing—but they've persisted for eleven years now. Although the burning is often tolerable during the day, the slightest thing can trigger an excruciating flare-up—a cool breeze across the skin, the brush of a shirtsleeve or a bedsheet. "Sometimes

I feel that my skin has been flayed and my flesh is exposed, and any touch is just very painful," he told me. "Sometimes I feel that there's an ice pick or a wasp sting. Sometimes I feel that I've been splattered with hot cooking oil."

For all that, the itch has been harder to endure. H. has developed calluses from the incessant scratching. "I find I am choosing itch relief over the pain that I am provoking by satisfying the itch," he said.

He has tried all sorts of treatments—medications, acupuncture, herbal remedies, lidocaine injections, electrical-stimulation therapy. But nothing really worked, and the condition forced him to retire in 2001. He now avoids leaving the house. He gives himself projects. Last year, he built a three-foot stone wall around his yard, slowly placing the stones by hand. But he spends much of his day, after his wife has left for work, alone in the house with their three cats, his shirt off and the heat turned up, trying to prevent a flare-up.

His neurologist introduced him to me, with his permission, as an example of someone with severe itching from a central rather than a peripheral cause. So one morning we sat in his living room trying to puzzle out what was going on. The sun streamed in through a big bay window. One of his cats, a scraggly brown tabby, curled up beside me on the couch. H. sat in an armchair in a baggy purple T-shirt he'd put on for my visit. He told me that he thought his problem was basically a "bad switch" in his neck where the tumor had been, a kind of loose wire sending false signals to his brain. But I told him about the increasing evidence that our sensory experiences are not sent to the brain but originate in it. When I got to the example of phantom-limb sensations, he perked up. The experiences of phantom-limb patients sounded familiar to him. When I mentioned that he might want to try the mirror-box treatment, he agreed. "I have a mirror upstairs," he said.

He brought a cheval glass down to the living room, and I had him stand with his chest against the side of it, so that his troublesome left arm was behind it and his normal right arm was in front. He tipped his head so that when he looked into the mirror the image of his right arm seemed to occupy the same position as his left arm. Then I had him wave his arms, his actual arms, as if he were conducting an orchestra.

The first thing he expressed was disappointment. "It isn't quite like looking at my left hand," he said. But then suddenly it was.

"Wow!" he said. "Now, this is odd."

After a moment or two, I noticed that he had stopped moving his left arm. Yet he reported that he still felt as if it were moving. What's more, the sensations in it had changed dramatically. For the first time in eleven years, he felt his left hand "snap" back to normal size. He felt the burning pain in his arm diminish. And the itch, too, was dulled.

"This is positively bizarre," he said.

He still felt the pain and the itch in his neck and shoulder, where the image in the mirror cut off. And, when he came away from the mirror, the aberrant sensations in his left arm returned. He began using the mirror a few times a day, for fifteen minutes or so at a stretch, and I checked in with him periodically.

"What's most dramatic is the change in the size of my hand," he says. After a couple of weeks, his hand returned to feeling normal in size all day long.

The mirror also provided the first effective treatment he has had for the flares of itch and pain that sporadically seize him. Where once he could do nothing but sit and wait for the torment to subside—it sometimes took an hour or more—he now just pulls out the mirror. "I've never had anything like this before," he said. "It's my magic mirror."

THERE HAVE BEEN OTHER, ISOLATED successes with mirror treatment. In Bath, England, several patients suffering from what is called complex regional pain syndrome—severe, disabling limb sensations of unknown cause—were reported to have experienced complete resolution after six weeks of mirror therapy. In California, mirror therapy helped stroke patients recover from a condition known as hemineglect, which produces something like the opposite of a phantom limb—these patients have a part of the body they no longer realize is theirs.

Such findings open up a fascinating prospect: perhaps many patients whom doctors treat as having a nerve injury or a disease have, instead, what might be called sensor syndromes. When your car's dashboard warning light keeps telling you that there is an engine failure, but the mechanics can't find anything wrong, the sensor itself may be the problem. This is no less true for human beings. Our sensations of pain, itch, nausea, and fatigue are normally protective. Unmoored from physical reality, however, they can become a nightmare: M., with her intractable itching, and H., with his constellation of strange symptoms—but perhaps also the hundreds of thousands of people in the United States alone who suffer from conditions like chronic back pain, fibromyalgia, chronic pelvic pain, tinnitus, temporomandibular joint disorder, or repetitive strain injury, where, typically, no amount of imaging, nerve testing, or surgery manages to uncover an anatomical explanation. Doctors have persisted in treating these conditions as nerve or tissue problems—engine failures, as it were. We get under the hood and remove this, replace that, snip some wires. Yet still the sensor keeps going off.

So we get frustrated. "There's nothing wrong," we'll insist. And, the next thing you know, we're treating the driver instead of the problem. We prescribe tranquillizers, antidepressants, escalating doses of narcotics. And the drugs often do make it easier for people to ignore the sensors, even if they are wired right into the brain. The mirror treatment, by contrast, targets the deranged sensor system itself. It essentially takes a misfiring sensor—a warning system functioning under an illusion that something is terribly wrong out in the world it monitors—and feeds it an alternate set of signals that calm it down. The new signals may even reset the sensor.

This may help explain, for example, the success of the advice that back specialists now commonly give. Work through the pain, they tell many of their patients, and, surprisingly often, the pain goes away. It had been a mystifying phenomenon. But the picture now seems clearer. Most chronic back pain starts as an acute back pain—say, after a fall. Usually, the pain subsides as the injury heals. But in some cases the pain sensors continue to light up long after the tissue damage is gone. In such instances, working through the pain may offer the brain contradictory feedback—a signal that ordinary activity does not, in fact, cause physical harm. And so the sensor resets.

This understanding of sensation points to an entire new array of potential treatments—based not on drugs or surgery but, instead, on the careful manipulation of our perceptions. Researchers at the University of Manchester, in England, have gone a step beyond mirrors and fashioned an immersive virtual-reality system for treating patients with phantom-limb pain. Detectors transpose movement of real limbs into a virtual world where patients feel they are actually moving, stretching, even playing a ballgame. So far, five patients have tried the system, and they have all experienced a reduction in pain. Whether those results will last has yet to be established. But the approach raises the possibility of designing similar systems to help patients with other sensor syndromes. How, one wonders, would someone with chronic back pain fare in a virtual world? The Manchester study suggests that there may be many ways to fight our phantoms.

I called Ramachandran to ask him about M.'s terrible itch. The sensation may be a phantom, but it's on her scalp, not in a limb, so it seemed unlikely that his mirror approach could do anything for her. He told me about an experiment in which he put ice-cold water in people's ears. This confuses the brain's position sensors, tricking subjects into thinking that their heads are moving, and in certain phantom-limb and stroke patients the illusion corrected their misperceptions, at least temporarily. Maybe this would help M., he said. He had another idea. If you take two mirrors and put them at right angles to each other, you will get a non-reversed mirror image. Looking in, the right half of your face appears on the left and the left half appears on the right. But unless you move, he said, your brain may not realize that the image is flipped.

"Now, suppose she looks in this mirror and scratches the left side of her head. No, wait—I'm thinking out loud here—suppose she looks and you have *someone else* touch the left side of her head. It'll look—maybe it'll feel—like you're touching the right side of her head." He let out an impish giggle. "Maybe this would make her itchy right scalp feel more normal." Maybe it would encourage her brain to make a different perceptual inference; maybe it would press reset. "Who knows?" he said.

It seemed worth a try.

SALLIE TISDALE

Twitchy

FROM *THE ANTIOCH REVIEW*

Sallie Tisdale has weak teeth and hates going to the dentist. But dealing with the pain and the anxiety about the pain offer her insights into the patients she sees as an oncological nurse.

WHEN I GO TO THE DENTIST, I TAKE MY IPOD, AND crank it up. The right music is important: big music, strong-flavored and complex, but not too sweet. Piano concertos don't work, and folk songs are all wrong. I listen to the Doors and Art of Noise and Talking Heads. Depending on how the visit goes, whatever music I choose will be tainted for months afterward—a bit dentisty, with the faint scent of nitrous.

I pull off the headphones when the assistant appears. Sharon is my favorite, a sassy, skinny blonde of a certain age who makes cracks under her breath. She shares the same bleak humor I bring to the chair. We chat a little, and then she puts the dark rubber nose mask on me, and the smell is the smell of emptiness, the smell of no smell at all. She wraps a bib around my chin and then gives me big black safety glasses, and finally I put the headphones back on. They prop my mouth open and stretch the rubber saliva dam across my tongue and hang the slurping suction tube on my lip and then I have a hot flash and start panting like a dog. I know I look like a drunk on Halloween, but it's a passing concern. Fashion's not my worry here.

"We're turning the nitrous on now," she says. I sink back, eyes half-closed, and turn the volume up a little more on *Light My Fire*. The empty smell is gone, replaced by a light breeze, and I take it in deeply.

This is one of those modern dental offices, the kind you find near suburban shopping malls. All the dentists are women, and the walls are pink and pale blue and light green, soothing pastel walls with big murals of tropical islands and country roads. There are skylights above the chairs and everyone wears colorful scrubs. The waiting room has a children's play area and lots of good magazines.

I watch a cloud cross the skylight.

Dr. Johnston appears at my side, and speaks to me. I mumble, *we've got to stop meeting like this,* and then her looming face drifts back and I'm humming to myself again.

She murmurs to Sharon, picks up a silver tool, prods a bit here and there.

I hear Sharon say, from a distance, "She's pretty twitchy today."

"Yeah—why don't you turn her up a little?"

I smile to myself. Yes, turn me up a little, I think. I watch a cloud cross the skylight. On nitrous, there is a vague numbness like a line drawn around the edges of my body. Sometimes I feel as though I am gently bobbing in the small eddy of a stream—moving, but going nowhere. I know what's happening, what happens next; I hear snippets of their chatting, something about weekend plans, and I resent it. I want them attending to *me* right now, but my resentment is distant, too, and fades away. I watch the sky, a cloud inflating in an interesting way, expanding into a yellow sheet and then darkening, and R. L. Burnside growls, "Baby done a baad, baaaad thing," and the yellow cloud is gone and all I see is blue.

watch   cloud cross   sky

I have weak teeth, and of course I blame my mother, who gave birth to me in 1957 when everyone smoked and drank martinis during pregnancy. My mother's pleasures were few enough—coffee, cigarettes, romance novels, and a bowl of ice cream every evening—but she enjoyed them deeply. A few months before she died of breast cancer, she said out of nowhere, "You know, there are two things I'd do differently in my life if I could do it over." And I'm looking at my father sacked out on the couch, and thinking, only two?

"I wouldn't start smoking"—and that hangs in the air a moment—"and I wouldn't let you kids eat so much sugar."

I'm not sure why the last mattered much to her then. Considering.

My brother and sister and I spent a lot of hours at the dentist, and more hours drinking Dr Pepper and eating Popsicles. We used to get sugar sandwiches as a treat—white sugar, margarine, and white bread—Wonder bread, which builds strong bodies all kinds of ways. Perhaps the Flintstones vitamins weren't the miracle cure the commercials led her to believe. The inadequate brushing from a childhood lived without much supervision took its toll. But I suspect this is something more buried and intractable—genetics or karma or destiny—something embedded in the whole of me. Either way, this is the result: countless fillings, several crowns, many teeth more false than not, and long afternoons in the chair beneath the skylight.

Nitrous oxide is a mysterious drug, though in part it is antinociceptive; that means it blocks some of the sensory perception of pain, just like Advil. But the lidocaine takes care of that. In fact, I don't mind the shot, the part most people dread the most. I don't need the nitrous to prevent pain. I need it for the fear of pain, for the fear of the drill, for the smell, the grinding noise, the vibration, the gestalt of a world I find vaguely terrifying far out of proportion to my experience. I discovered nitrous many years ago, thanks to a dentist who knew I wouldn't get my teeth fixed otherwise. Unlike my other fears—plane crashes, apologizing—this one has not lessened in time. Everything changed with nitrous; I still hated to go, but I could go.

Why nitrous works this way, no one really knows. Nitrous reduces anxiety and makes people feel happy, and the medical literature invariably describes the effect as "pleasant." People report sensations of tingling, vibration, throbbing or droning sounds, warmth, a sense of being heavy or floating. Patients on nitrous are suggestible, accommodating, with "an indifference to surroundings." Indifference is, above all, what I lack in the chair.

I am a connoisseur of dosing. Dentists typically use a 20-percent mixture; I need about 30 percent to get through a procedure, a dose that causes some people to lose consciousness. For me, it's just right. Over the course of an hour, I feel myself fading slightly in and out of the room, separate from the complicated maneuvers going on in my mouth. I am a little removed, but not gone; the tension remains. Now and then I am reminded to unclench my jaw, relax the hands which have somehow become tightly wrapped around the arms of the chair. We have signals. Up, I raise my thumb. Thumbs up, for more gas.

People still call nitrous oxide the laughing gas. I don't get the giggles like some people do, but I make some very funny jokes—profound, meaningful jokes with a subtle poignancy and the hint of tragedy that brings classic humor to life. I never forget that I've taken a drug, but somehow the idea that my funny jokes, the funniest jokes I've ever made, are not really very funny and are entirely the product of chemistry, makes them funnier still. There is some vast and deeper humor here, a whistling past the graveyard that lightens the burden of my fear.

Dr. Johnston has been my dentist for a long time. We've gotten to know each other over the years, and have a friendly relationship. One day, as I am fading in and out of the room, she suggests that we take care of a filling at the same time she works on the crown. "Get it over with," she says.

"Yeah," I smile. Long pause. "Why don't you do a pelvic while you're at it and cover all the bases?" I think this is very funny, and start chuckling.

She just thinks *I'm* funny right now, and rolls with it. "I'm afraid I'd get arrested for that," she says.

"Oh, we should just get married, we love each other so much," I say, apropos of who knows what.

"That's illegal, too," she answers, and after a few more laughs, I go back to the clouds, and she goes back to work.

I'm sitting in the park on a sunny day, reading my book and eating a bagel. And crack. A *bagel.* A bit of my tooth falls into my hand. And here I go again; I know what's next. I know the whole story, each chapter, and the long arcs of tension and release to come.

I come in to have the impression made for another crown. Dr. Johnston is gone, and this is my first appointment with Dr. Bennett. Dr. Johnston is brisk and practical. Dr. B is padded and cheerful, and seems willing to wait as long as necessary for me to get comfortable. I'm nervous today—I'm always nervous, but she's a stranger to me, and so I decide to tell Dr. B about Grandpa Doc.

Large sections of my childhood reel off in my head like a film noir starring Gloria Grahame—all sharp shadows and lurching camera moves. Grandpa Doc wasn't really my grandfather. My grandfather, the rumor goes, killed himself, but no one ever talked about it. To this day I'm not sure what really happened; there's no one left to ask. Doc was my grandmother's third husband—a tall, thin, bald dentist who loved to fish and hunt. I saw him in my parents' living room, holding a highball and laughing, and I saw him in his office. He didn't like children much, but as a favor, he did our dental work for free on the weekends, when his office was closed.

Childhood trips to the dentist are one of the dream sequences in my dark B movie, full of quick cuts and freeze frames: cold, dim office; echoing voices; white walls; bright metal trays. I'm alone in a high, high chair. The images are vague but immediate, lingering the way dreams do. I am six years old, under fluorescent light, and looming over me is a tall, thin, bald old man in a high-collared white coat.

He is not smiling.

I asked my brother recently what he remembered of these visits, and he answered with the same bleak smile I bring to the chair today.

"I had the pleasure of that office many times," he said. "My youthful oral hygiene was more like lowgiene."

"Were you scared of Grandpa Doc?" I asked. My brother is a little older than I; we were partners as children, friends to each other along the way. He is the only person I can ask about things like this.

"I don't remember being afraid of *him*," he said. "Just of the upcoming procedures. The needle, the awful taste of the Novocain that trickled out, the sound, and of course the swell stench of ground enamel."

When I think about Grandpa Doc, I don't remember the drill. I don't remember shots. I don't even remember pain. Everything is much more fragmented than that. The tile floor, a long way down—the big cold vinyl arms of the chair—the bald man bending way down, face close to mine—that's what I remember. What I remember is dread. Dr. B says, "Did your mother really send you there *alone*?" She did—she was sitting out in the waiting room with my brother and sister, waiting their turns.

"He was my *grandpa*!" I say. "You're supposed to like your grandpa." I want to defend my mother, who did the best she could. "Besides," I add, "I don't think I ever told her I was scared."

There were a lot of things I never told my mother.

Those first visits to Grandpa Doc became visits to a nice young dentist in a sunny office up the street from our house, and in time, led to my steadfast refusal to see a dentist at all. Do I have a post-traumatic stress disorder, as one helpful friend suggested? No, as long as I'm not actually at the dentist, there's no problem. Am I phobic? Phobias are usually a bit more occult, and this isn't as hard to explain as a fear of ants or peanut butter. I think this is as simple as it sounds: a lonesome and somewhat secretive child made to do something hard and never telling how hard it was and holding on until she was old enough to say, no more. Telling it to Dr. B in all the pink light, I feel a little silly. But when I get the stupid nose mask on and take a deep breath, all I feel is relief.

I GO FOR A NEW kind of porcelain crown, because this time the tooth I've broken is in the front. It costs about the same, and this new kind of crown is supposed to be easier; it requires only one appointment instead of two. Dr. Johnston doesn't do this procedure, so that means seeing another new dentist, Dr. Fischl, a slim young blonde who looks about twenty-one years old. Like all of the dentists, at first she doesn't understand why I need nitrous for such a benign procedure.

"There's hardly any drilling at all!" she exclaims.

I reach for the mask.

I watch the light above me, the bright color in the sky. I know where I am, and what's happening. "I'm going to numb you now," she says, and I feel the weird push of the shot into my gum, the slowly spreading fog that blanks out cheek, lip, tongue, my lower eyelid. But she goes for the 20 percent, and except for a few brief moments, I never have enough nitrous. I strain for it, as though in chains. Then I find out that this easy new kind of crown involves a slow, vibrating drill that rattles the inside of my skull like a hammer and makes me shiver from scalp to toes, and that the one appointment is a very long one.

When everything is done but the last fitting on my new porcelain crown, I can finally relax a little bit, knowing we're almost finished. But Dr. Fischl doesn't know me. She turns off the nitrous and switches to oxygen, without telling me—before the crown is polished. After all, this is when they typically turn off the nitrous for everyone else. It hasn't been enough, but a lot more than nothing. I don't know what's going on; I only know that I'm coming back while the work goes on—polishing and grinding, tugging and pushing, the buzzing and the *smells*, and I've shrunk into a single cramped shell and the minutes stretch very slowly out and on and on as I twist and flinch inside like a cat stuffed into a small box.

When I'm finally able to make a sensible complaint, the assistant tells me, "But the work is just on the crown, not the tooth!"

"Look," I say. "Look. You can't turn it down. You can't, not until we're done, not if there's any kind of drilling or polishing or *anything*."

"But it's not on your tooth, it's on the crown," she says again.

"*No,*" I answer. "No—no vibration, the sound, the smell, I don't know, I *can't*."

She looks at me for a moment, and I guess she sees something in my face, because she just nods. "Okay," she says. "Okay. I'll make a note."

ONE DAY, ONE OF THE dentists told me not to exhale through my mouth. I had thought I should do just that, in order not to inhale my own carbon dioxide. (My sneaky hope being, of course, that I could get a bigger dose of nitrous that way; I thought I might be getting away with something.)

"The nitrous off-gases when you breathe," she said. "Breathe through your nose, so we don't get any of the nitrous ourselves." Then she paused. "Not that it isn't fun."

I work part-time as an oncology nurse; part of my job is to cause pain, and a bigger part is to treat it, which means to witness it, ask about it, talk about it, listen to all the things people want to say about it. It means acknowledging the almost infinite number of ways we shame ourselves. A lot of people are ashamed of their pain, and most are a little ashamed that they want it to go away.

Somehow, and this is a story for another day, people come to believe that they should feel pain, that they deserve it or must accept it. I completely agree in principle: we harm ourselves in endless ways with avoidance and denial, by resisting the facts of the life before us. I think life is best lived head-on, leaning forward into what comes: into loss, gain, change. Facing things squarely—this is the straightest way through, the way of least suffering, and most joy. But I don't extend this to physical pain that can be treated. There is enough waiting for us that can't be helped.

I HAVE A FRIEND WHO says he looks forward to going to the dentist, because it's a challenge—he likes to find out if all his careful flossing has worked. Okay—there's a bit of the Boy Scout in that, but I get his motivation. I know two people who don't use lidocaine. One has an allergy. Another—well, she believes that life is suffering, and that we should feel it, that we shouldn't flinch, that resistance is futile. I agree, up to a point; but when the shot is there and a few billion people would be grateful for it, not resisting may be the irrational choice. And I think she has some twisted idea of what it means to be strong; she wants others to watch her feel it and be impressed by her power to handle the pain, which is at least a little insane, I think. But so am I, on the other side of things. Like most counselors, I need my own advice; in the dentist's chair, I am like a flinching boxer past his prime, dancing around the ring with the young contender. I try to face it squarely, but what I face is how to make it through the fear at all.

I want the dentists to reassure me, tell me that my cowardice is understandable, or at least not that uncommon. But I know I am way out on one end of this continuum. I know the pain can be treated, but the existential terror of pain known from the past becomes a new pain today. I am curious about almost everything in the world but I am not curious about dentistry, about the battlefield of my mouth. Curiosity requires the willingness to engage, to know. I actually conquered a fear of flying; I talked myself out of it by learning and rational discussion and a few judicious doses of Xanax and Inderal. I can't do that in the dentist's chair; I can't begin to engage. This is the plain fact: I'm not as afraid of being afraid, or of being ashamed of being afraid, as I am afraid of the dentist.

A long time ago, I worked in an outpatient surgical clinic. We sometimes used nitrous on our patients. At the end of a long day, one of the doctors, another nurse, and I were sitting around the back room. We were all feeling the stress of too many patients, too many hours. The doctor reached out and grabbed the mask on the nitrous cylinder, spun the knob, and took a couple of whiffs. He smiled and handed it to me. Straight from his hand to mine, like a prescription; I took the mask and a few quick breaths and felt a little light-headed and then guilty. My drug reference lists "euphoria" as an "adverse effect" of narcotics, because feeling good is not the intended purpose of narcotics. When I took that single whiff of nitrous oxide, I felt as if I were cheating in some big way—cheating because I didn't "need" it, because I wasn't sick, because it felt good. Because it was fun. I never cease to wonder at the vast injury we have done ourselves this way—that we fear feeling pleasure means not paying for our sins.

THERE HAS BEEN THE ONE root canal. I dreaded root canals the way other people dread public speaking or cancer. I go for a consultation first. Just to talk. The endodontist has a large binder of helpful full-color drawings telling me exactly what he is going to do down in my tooth root with his gutta percha and drill and strange tools. He looks at my X-rays and says, pointing, "I think you'll need a root canal on this other tooth, as well." He pauses. "We've got time, we can do it right now."

The tears start rolling down my face. The whole world has shrunk to the size of a root canal, or the root of my tooth is the size of the world, I can no longer tell. I look at him and see that he is embarrassed.

A little annoyed with me, with my babyish tears.

Even now, I'm surprised at the lack of sympathy. I am periodically reminded by my patients that what is normal to me (the sounds, the smells) is a strange and frightening world to them. I'm not the typical dental patient, and I have to remind them of that. This is very far from normal to me. I don't like the looks of the tools; I quite hate the sounds coming from the next cubicle, and I'm not in what would be called the proper frame of mind for learning. I had asked the receptionist about sedation as soon as I arrived—real sedation, a drug called Versed that causes amnesia. It's used for painful procedures like colonoscopies and bronchoscopies, and I've seen time and again how easily patients get through difficult hours this way.

"It costs $500," she tells me, and so that's that. It's lidocaine and nitrous, or nothing.

When I finally arrive for the procedure, a few weeks later, I find out that one of the assistants is the woman who was my best friend for a while in high school, when we prowled the streets of our small town in a plague of dissatisfaction and restless urges. She makes cheerful small talk while I take the chair, the stress vibrating through me. She gets me started on the mask, chatting about my siblings and the weather and her apartment and her boyfriend and her car and who cares and I don't care and I watch the cloud passing across the window and after a while stop answering and after a while she leaves, and I think to myself that no one's feelings will be hurt because we can blame it on the nitrous. On nitrous, all is forgiven.

The root canal didn't hurt more than a filling or a crown. Pain—the physical pain is covered. The numbness required was deep and extensive, spreading through my eyeballs, far up into my nostrils and scalp, into the canals of my ears, and it lingered for many hours. The procedure was long, so long, on and on. And I am cursed with imagination, and wish I'd never seen the binder with its big color pictures. It filled my dreams.

TIME TO HAVE ANOTHER PERMANENT crown cemented on, a gold one on a back tooth. This is a procedure many people do without lidocaine. In the days before an appointment, I might go through a rainbow of experience: stress and anger, an undirected vague resentment, self-pity, more impotent anger. As it approaches, anxiety begins to crowd out everything else, anxiety felt in the palms and stomach. The morning of the appointment, the secretary calls and warns me there is something wrong with the nitrous setup and it won't be available today. My tooth is hurting, I'm sick of it all, I want to be done. This one tooth has probably been worked on eight times or more in my life. I hang up and a small, bleak terror envelops me and I'm crying, lying on my bed in the dark. *I can't I can't I can't I can't.* And I don't; I make another appointment.

When I get there a few days later, I try to explain to Dr. Bennett. It has been so long since I've had dental work without nitrous. "I kind of fell apart," I say. "I really am a strong person," I say. "I'm not scared of many things. Really, I'm pretty good in most places. Really." I imagine my pioneer ancestors facing the pliers with only a bottle of whiskey. I imagine the sidewalk dentists in India. None of that helps; gratitude is a crappy antidote for terror. I look up and back at Dr. B, where she sits behind me, and say, "You're my mommy now." She smiles, but doesn't laugh. We're all five years old inside. I've told patients this a thousand times, and I tell myself the same: we are all children, we are all vulnerable, we are all helpless in each other's hands.

She hands me the mask. The pink walls turn orange, the light expands, a tingling, cottony sensation surrounds me. I turn up the music, and David Byrne is crooning to me: *Hold tight—we're in for nasty weather.* Pinching. Pushing and pulling, clamping and scraping. Empty skylight. No cloud.

I realize that I am alone. There is no one near me. Perhaps they have forgotten me. Perhaps they went to lunch, or home for the weekend. I feel a rush of self-pity. *I'm just going to sit here, all alone*, I think. My pity slowly drifts across the skylight with everything else, finally, like everything else, fading away.

ANNIE MURPHY PAUL

The First Ache

FROM *THE NEW YORK TIMES MAGAZINE*

For the last quarter century, consensus has emerged among medical researchers that fetuses and newborn infants can experience pain—whereas, before, such claims were generally dismissed. But as Annie Murphy Paul shows, while these findings may alter our conventional wisdom about the nature of consciousness, they also add to the debate over abortion.

TWENTY-FIVE YEARS AGO, WHEN KANWALJEET ANAND was a medical resident in a neonatal intensive care unit, his tiny patients, many of them preterm infants, were often wheeled out of the ward and into an operating room. He soon learned what to expect on their return. The babies came back in terrible shape: their skin was gray, their breathing shallow, their pulses weak.

Anand spent hours stabilizing their vital signs, increasing their oxygen supply and administering insulin to balance their blood sugar.

"What's going on in there to make these babies so stressed?" Anand wondered. Breaking with hospital practice, he wrangled permission to follow his patients into the O.R. "That's when I discovered that the babies were not getting anesthesia," he recalled recently. Infants undergoing major surgery were receiving only a paralytic to keep them still. Anand's encounter with this practice occurred at John Radcliffe Hospital in Oxford, England, but it was common almost everywhere. Doctors were convinced that newborns' nervous systems were too immature to sense pain, and that the dangers of anesthesia exceeded any potential benefits.

Anand resolved to find out if this was true. In a series of clinical trials, he demonstrated that operations performed under minimal or no anesthesia produced a "massive stress response" in newborn babies, releasing a flood of fight-or-flight hormones like adrenaline and cortisol. Potent anesthesia, he found, could significantly reduce this reaction. Babies who were put under during an operation had lower stress-hormone levels, more stable breathing and blood-sugar readings and fewer postoperative complications. Anesthesia even made them more likely to survive. Anand showed that when pain relief was provided during and after heart operations on newborns, the mortality rate dropped from around 25 percent to less than 10 percent. These were extraordinary results, and they helped change the way medicine is practiced. Today, adequate pain relief for even the youngest infants is the standard of care, and the treatment that so concerned Anand two decades ago would now be considered a violation of medical ethics.

But Anand was not through with making observations. As NICU technology improved, the preterm infants he cared for grew younger and younger—with gestational ages of 24 weeks, 23, 22—and he noticed that even the most premature babies grimaced when pricked by a needle. "So I said to myself, Could it be that this pain system is developed and functional before the baby is born?" he told me in the fall. It was not an abstract question: fetuses as well as newborns may now go under the knife. Once highly experimental, fetal surgery—to remove lung tumors, clear blocked urinary tracts, repair malformed diaphragms—is a frequent occurrence at a half-dozen fetal treatment centers around the country, and could soon become standard care for some conditions diagnosed prenatally like spina bifida. Whether the fetus feels pain is a question that matters to the doctor wielding the scalpel.

And it matters, of course, for the practice of abortion. Over the past four years, anti-abortion groups have turned fetal pain into a new front in their battle to restrict or ban abortion. Anti-abortion politicians have drafted laws requiring doctors to tell patients seeking abortions that a fetus can feel pain and to offer the fetus anesthesia; such legislation has already passed in five states. Anand says he does not oppose abortion in all circumstances but says decisions should be made on a case-by-case basis. Nonetheless, much of the activists' and lawmakers' most powerful rhetoric on fetal pain is borrowed from Anand himself.

Known to all as Sunny, Anand is a soft-spoken man who wears the turban and beard of his Sikh faith. Now a professor at the University of Arkansas for Medical Sciences and a pediatrician at the Arkansas Children's Hospital in Little Rock, he emphasizes that he approaches the question of fetal pain as a scientist: "I eat my best hypotheses for breakfast," he says, referring to the promising leads he has discarded when research failed to bear them out. New evidence, however, has persuaded him that fetuses can feel pain by 20 weeks gestation (that is, halfway through a full-term pregnancy) and possibly earlier. As Anand raised awareness about pain in infants, he is now bringing attention to what he calls "signals from the beginnings of pain."

But these signals are more ambiguous than those he spotted in newborn babies and far more controversial in their implications. Even as some research suggests that fetuses can feel pain as preterm babies do, other evidence indicates that they are anatomically, biochemically and psychologically distinct from babies in ways that make the experience of pain unlikely. The truth about fetal pain can seem as murky as an image on an ultrasound screen, a glimpse of a creature at once recognizably human and uncomfortably strange.

If the notion that newborns are incapable of feeling pain was once widespread among doctors, a comparable assumption about fetuses was even more entrenched. Nicholas Fisk is a fetal-medicine specialist and director of the University of Queensland Center for Clinical Research in Australia. For years, he says, "I would be doing a procedure to a fetus, and the mother would ask me, 'Does my baby feel pain?' The traditional, knee-jerk reaction was, 'No, of course not.'" But research in Fisk's laboratory (then at Imperial College in London) was making him uneasy about that answer. It showed that fetuses as young as 18 weeks react to an invasive procedure with a spike in stress hormones and a shunting of blood flow toward the brain—a strategy, also seen in infants and adults, to protect a vital organ from threat. Then Fisk carried out a study that closely resembled Anand's pioneering research, using fetuses rather than newborns as his subjects. He selected 45 fetuses that required a potentially painful blood transfusion, giving one-third of them an injection of the potent painkiller fentanyl. As with Anand's experiments, the results were striking: in fetuses that received the analgesic, the production of stress hormones was halved, and the pattern of blood flow remained normal.

Fisk says he believes that his findings provide suggestive evidence of fetal pain—perhaps the best evidence we'll get. Pain, he notes, is a subjective phenomenon; in adults and older children, doctors measure it by asking patients to describe what they feel. ("On a scale of 0 to 10, how would you rate your current level of pain?") To be certain that his fetal patients feel pain, Fisk says, "I would need one of them to come up to me at the age of 6 or 7 and say, 'Excuse me, Doctor, that bloody hurt, what you did to me!'" In the absence of such first-person testimony, he concludes, it's "better to err on the safe side" and assume that the fetus can feel pain starting around 20 to 24 weeks.

Blood transfusions are actually among the least invasive medical procedures performed on fetuses. More intrusive is endoscopic fetal surgery, in which surgeons manipulate a joystick-like instrument while watching the fetus on an ultrasound screen. Most invasive of all is open fetal surgery, in which a pregnant woman's uterus is cut open and the fetus exposed. Ray Paschall, an anesthesiologist at Vanderbilt Medical Center in Nashville, remembers one of the first times he provided anesthesia to the mother and minimally to the fetus in an open fetal operation, more than 10 years ago. When the surgeon lowered his scalpel to the 25-week-old fetus, Paschall saw the tiny figure recoil in what looked to him like pain. A few months later, he watched another fetus, this one 23 weeks old, flinch at the touch of the instrument. That was enough for Paschall. In consultation with the hospital's pediatric pain specialist, "I tremendously upped the dose of anesthetic to make sure that wouldn't happen again," he says. In the more than 200 operations he has assisted in since then, not a single fetus has drawn back from the knife. "I don't care how primitive the reaction is, it's still a human reaction," Paschall says. "And I don't believe it's right. I don't want them to feel pain."

But whether pain is being felt is open to question. Mark Rosen was the anesthesiologist at the very first open fetal operation, performed in 1981 at the University of California, San Francisco, Medical Center, and the fetal anesthesia protocols he pioneered are now followed by his peers all over the world. Indeed, Rosen may have done more to prevent fetal pain than anyone else alive—except that he doesn't believe that fetal pain exists. Research has persuaded him that before a point relatively late in pregnancy, the fetus is unable to perceive pain.

Rosen provides anesthesia for a number of other important reasons, he explains, including rendering the pregnant woman unconscious and preventing her uterus from contracting and setting off dangerous bleeding or early labor. Another purpose of anesthesia is to immobilize the fetus during surgery, and indeed, the drugs Rosen supplies to the pregnant woman do cross the placenta to reach the fetus. Relief of fetal pain, however, is not among his objectives. "I have every reason to want to believe that the fetus feels pain, that I've been treating pain all these years," says Rosen, who is intense and a bit prickly. "But if you look at the evidence, it's hard to conclude that that's true."

Rosen's own hard look at the evidence came a few years ago, when he and a handful of other doctors at U.C.S.F. pulled together more than 2,000 articles from medical journals, weighing the accumulated evidence for and against fetal pain. They published the results in *The Journal of the American Medical Association* in 2005. "Pain perception probably does not function before the third trimester," concluded Rosen, the review's senior author. The capacity to feel pain, he proposed, emerges around 29 to 30 weeks gestational age, or about two and a half months before a full-term baby is born. Before that time, he asserted, the fetus's higher pain pathways are not yet fully developed and functional.

What about a fetus that draws back at the touch of a scalpel? Rosen says that, at least early on, this movement is a reflex, like a leg that jerks when tapped by a doctor's rubber mallet. Likewise, the release of stress hormones doesn't necessarily indicate the experience of pain; stress hormones are also elevated, for example, in the bodies of brain-dead patients during organ harvesting. In order for pain to be felt, he maintains, the pain signal must be able to travel from receptors located all over the body, to the spinal cord, up through the brain's thalamus and finally into the cerebral cortex. The last leap to the cortex is crucial, because this wrinkly top layer of the brain is believed to be the organ of consciousness, the generator of awareness of ourselves and things not ourselves (like a surgeon's knife). Before nerve fibers extending from the thalamus have penetrated the cortex—connections that are not made until the beginning of the third trimester—there can be no consciousness and therefore no experience of pain.

Sunny Anand reacted strongly, even angrily, to the article's conclusions. Rosen and his colleagues have "stuck their hands into a hornet's nest," Anand said at the time. "This is going to inflame a lot of scientists who are very, very concerned and are far more knowledgeable in this area than the authors appear to be. This is not the last word—definitely not." Anand acknowledges that the cerebral cortex is not fully developed in the fetus until late in gestation. What is up and running, he points out, is a structure called the subplate zone, which some scientists believe may be capable of processing pain signals. A kind of holding station for developing nerve cells, which eventually melds into the mature brain, the subplate zone becomes operational at about 17 weeks. The fetus's undeveloped state, in other words, may not preclude it from feeling pain. In fact, its immature physiology may well make it more sensitive to pain, not less: the body's mechanisms for inhibiting pain and making it more bearable do not become active until after birth.

The fetus is not a "little adult," Anand says, and we shouldn't expect it to look or act like one. Rather, it's a singular being with a life of the senses that is different, but no less real, than our own.

THE SAME MIGHT BE SAID of the five children who were captured on video by a Swedish neuroscientist named Bjorn Merker on a trip to Disney World a few years ago. The youngsters, ages 1 to 5, are shown smiling, laughing, fussing, crying; they appear alert and aware of what is going on around them. Yet each of these children was born essentially without a cerebral cortex. The condition is called hydranencephaly, in which the brain stem is preserved but the upper hemispheres are largely missing and replaced by fluid.

Merker (who has held positions at universities in Sweden and the United States but is currently unaffiliated) became interested in these children as the living embodiment of a scientific puzzle: where consciousness originates. He joined an online self-help group for the parents of children with hydranencephaly and read through thousands of e-mail messages, saving many that described incidents in which the children seemed to demonstrate awareness. In October

2004, he accompanied the five on the trip to Disney World, part of an annual get-together for families affected by the condition. Merker included his observations of these children in an article, published last year in the journal *Behavioral and Brain Sciences*, proposing that the brain stem is capable of supporting a preliminary kind of awareness on its own. "The tacit consensus concerning the cerebral cortex as the 'organ of consciousness,'" Merker wrote, may "have been reached prematurely, and may in fact be seriously in error."

Merker's much-discussed article was accompanied by more than two dozen commentaries by prominent researchers. Many noted that if Merker is correct, it could alter our understanding of how normal brains work and could change our treatment of those who are now believed to be insensible to pain because of an absent or damaged cortex. For example, the decision to end the life of a patient in a persistent vegetative state might be carried out with a fast-acting drug, suggested Marshall Devor, a biologist at the Center for Research on Pain at Hebrew University in Jerusalem. Devor wrote that such a course would be more humane than the weeks of potentially painful starvation that follows the disconnection of a feeding tube (though as a form of active euthanasia it would be illegal in the United States and most other countries). The possibility of consciousness without a cortex may also influence our opinion of what a fetus can feel. Like the subplate zone, the brain stem is active in the fetus far earlier than the cerebral cortex is, and if it can support consciousness, it can support the experience of pain. While Mark Rosen is skeptical, Anand praises Merker's work as a "missing link" that could complete the case for fetal pain.

But anatomy is not the whole story. In the fetus, especially, we can't deduce the presence or absence of consciousness from its anatomical development alone; we must also consider the peculiar environment in which fetuses live. David Mellor, the founding director of the Animal Welfare Science and Bioethics Center at Massey University in New Zealand, says he was prompted to consider the role of fetal surroundings in graduate school. "Have you ever wondered," one visiting professor asked, "why a colt doesn't get up and gallop around inside the mare?" After all, a horse only minutes old is already able to hobble around the barnyard. The answer, as Mellor reported in an influential review published in 2005, is that biochemicals produced by the placenta and fetus have a sedating and even an anesthetizing effect on the fetus (both equine and human). This fetal cocktail includes adenosine, which suppresses brain activity; pregnanolone, which relieves pain; and prostaglandin D_2, which induces sleep—"pretty potent stuff," he says.

Combined with the warmth and buoyancy of the womb, this brew lulls the fetus into a near-continuous slumber, rendering it effectively unconscious no matter what the state of its anatomy. Even the starts and kicks felt by a pregnant woman, he says, are reflex movements that go on in a fetus's sleep. While we don't know if the intense stimulation of surgery would wake it up, Mellor notes that when faced with other potential threats, like an acute shortage of oxygen, the fetus does not rouse itself but rather shuts down more completely in an attempt to conserve energy and promote survival. This is markedly different from the reaction of an infant, who will thrash about in an effort to dislodge whatever is blocking its airway. "A fetus," Mellor says, "is not a baby who just hasn't been born yet."

Even birth may not inaugurate the ability to feel pain, according to Stuart Derbyshire, a psychologist at the University of Birmingham in Britain. Derbyshire is a prolific commentator on the subject and an energetic provocateur. In milder moods, he has described the notion of fetal pain as a "fallacy"; when goaded by his critics' "lazy" thinking, he has pronounced it a "moral blunder" and "a shoddy, sentimental argument."

For all his vehemence in print, Derbyshire is affable in conversation, explaining that his laboratory research on the neurological basis of pain in adults led him to the matter of what fetuses feel: "For me, it's an interesting test case of what we know about pain. It's a great application of theory, basically." The theory, in this case, is that the experience of pain has to be learned—and the fetus, lacking language or interactions with caregivers, has no chance of learning it. In place of distinct emotions, it experiences a blur of sensations, a condition Derbyshire has likened to looking at "a vast TV screen with all of the world's information upon it from a distance of one inch; a great buzzing mass of meaningless information," he writes. "Before a symbolic system such as language, an individual will not know that something in front of them is large or small, hot or cold, red or green"—or, Derbyshire argues, painful or pleasant.

He finds "outrageous" the suggestion that the fetus feels anything like the pain that an older child or an adult experiences. "A fetus is biologically human, of course," he says. "It isn't a cow. But it's not yet psychologically human." That is a status not bestowed at conception but earned with each connection made and word spoken. Following this logic to its conclusion, Derbyshire has declared that babies cannot feel pain until they are 1 year old. His claim has become notorious in pain-research circles, and even Derbyshire says he thinks he may have overstepped. "I sometimes regret that I pushed it out quite that far," he concedes. "But really, who knows when the light finally switches on?"

IN FACT, "THERE MAY NOT be a single moment when consciousness, or the potential to experience pain, is turned on," Nicholas Fisk wrote with Vivette Glover, a colleague at Imperial College, in a volume on early pain edited by Anand. "It may come on gradually, like a dimmer switch." It appears that this slow dawning begins in the womb and continues even after birth. So where do we draw the line? When does a release of stress hormones turn into a grimace of genuine pain?

Recent research provides a potentially urgent reason to ask this question. It shows that pain may leave a lasting, even lifelong, imprint on the developing nervous system. For adults, pain is usually a passing sensation, to be waited out or medicated away. Infants, and perhaps fetuses, may do something different with pain: some research suggests they take it into their bodies, making it part of their fast-branching neural networks, part of their flesh and blood.

Anna Taddio, a pain specialist at the Hospital for Sick Children in Toronto, noticed more than a decade ago that the male infants she treated seemed more sensitive to pain than their female counterparts. This discrepancy, she reasoned, could be due to sex hormones, to anatomical differences—or to a painful event experienced by many boys: circumcision. In a study of 87 baby boys, Taddio found that those who had been circumcised soon after birth reacted more strongly and cried for longer than uncircumcised boys when they received a vaccination shot four to six months later. Among the circumcised boys, those who had received an analgesic cream at the time of the surgery cried less while getting the immunization than those circumcised without pain relief.

Taddio concluded that a single painful event could produce effects lasting for months, and perhaps much longer. "When we do something to a baby that is not an expected part of its normal development, especially at a very early stage, we may actually change the way the nervous system is wired," she says. Early encounters with pain may alter the threshold at which pain is felt later on, making a child hypersensitive to pain—or, alternatively, dangerously indifferent to it. Lasting effects might also include emotional and behavioral problems like anxiety and depression, even learning disabilities (though these findings are far more tentative).

Do such long-term effects apply to fetuses? They may well, especially since pain experienced in the womb would be even more anomalous than pain encountered soon after birth. Moreover, the ability to feel pain may not need to be present in order for "noxious stimulation"—like a surgeon's incision—to do harm to the fetal nervous system. This possibility has led some to venture an early end to the debate over fetal pain. Marc Van de Velde, an anesthesiologist and pain expert at University Hospitals Gasthuisberg in Leuven, Belgium, says: "We know that the fetus experiences a stress reaction, and we know that this stress reaction may have long-term consequences—so we need to treat the reaction as well as we can. Whether or not we call it pain is, to me, irrelevant."

* * *

BUT THE QUESTION OF FETAL pain is not irrelevant when applied to abortion. On April 4, 2004, Sunny Anand took the stand in a courtroom in Lincoln, Neb., to testify as an expert witness in the case of *Carhart* v. *Ashcroft*. This was one of three federal trials held to determine the constitutionality of the ban on a procedure called intact dilation and extraction by doctors and partial-birth abortion by anti-abortion groups. Anand was asked whether a fetus would feel pain during such a procedure. "If the fetus is beyond 20 weeks of gestation, I would assume that there will be pain caused to the fetus," he said. "And I believe it will be severe and excruciating pain."

After listening to Anand's testimony and that of doctors opposing the law, Judge Richard G. Kopf declared in his opinion that it was impossible for him to decide whether a "fetus suffers pain as humans suffer pain." He ruled the law unconstitutional on other grounds. But the ban was ultimately upheld by the U.S. Supreme Court, and Anand's statements, which he repeated at the two other trials, helped clear the way for legislation aimed specifically at fetal pain. The following month, Sam Brownback, Republican of Kansas, presented to the Senate the Unborn Child Pain Awareness Act, requiring doctors to tell women seeking abortions at 20 weeks or later that their fetuses can feel pain and to offer anesthesia "administered directly to the pain-capable unborn child." The bill did not pass, but Brownback continues to introduce it each year. Anand's testimony also inspired efforts at the state level. Over the past two years, similar bills have been introduced in 25 states, and in 5—Arkansas, Georgia, Louisiana, Minnesota and Oklahoma—they have become law. In addition, state-issued abortion-counseling materials in Alaska, South Dakota and Texas now make mention of fetal pain.

In the push to pass fetal-pain legislation, Anand's name has been invoked at every turn; he has become a favorite expert of the anti-abortion movement precisely because of his credentials. "This Oxford- and Harvard-trained neonatal pediatrician had some jarring testimony about the subject of fetal pain," announced the Republican congressman Mike Pence to the House of Representatives in 2004, "and it is truly made more astonishing when one considers the fact that Dr. Anand is not a stereotypical Bible-thumping pro-lifer." Anand maintains that doctors performing abortions at 20 weeks or later should take steps to prevent or relieve fetal pain. But it is clear that many of the anti-abortion activists who quote him have something more sweeping in mind: changing perceptions of the fetus. In several states, for example, information about fetal pain is provided to all women seeking abortions, including those whose fetuses are so immature that there is no evidence of the existence of even a stress response. "By personifying the fetus, they're trying to steer the woman's decision away from abortion," says Elizabeth Nash, a public-policy associate at the Guttmacher Institute, a reproductive-rights group.

Another, perhaps intended, effect of fetal-pain laws may be to make abortions harder to obtain. Laura Myers, an anesthesia researcher at Children's Hospital Boston and Harvard Medical School who analyzed the Unborn Child Protection Act for the abortion-rights organization Physicians for Reproductive Choice and Health, concluded that abortion clinics do not have the equipment or expertise to supply fetal anesthesia. "The handful of centers that perform fetal surgery are the only ones with any experience delivering anesthesia directly to the fetus," Myers says. "The bill makes a promise that the medical community can't fulfill." Even these specialized centers have no experience providing fetal anesthesia during an abortion; such a procedure would be experimental and would inevitably carry risks for the woman, including infection and uncontrolled bleeding.

In his speeches about fetal pain, Senator Brownback often asks why a fetus undergoing surgery receives anesthesia but not a fetus "who is undergoing the life-terminating surgery of an abortion." Mark Rosen rejects the analogy. "Fetal surgery is a different circumstance than abortion," he says, pointing out that none of the objectives of anesthesia for fetal surgery—relaxing the uterus, for example—apply to the termination of pregnancy. That includes an objective identified just recently: preventing possible long-term damage. For the fetus that is to be aborted, there is no long term. And if there is no pain, as Rosen maintains, then there is no cause to put the woman's health at risk.

Rosen sees no contradiction in his position, only a necessary complexity. When he was in medical school, he says, he worked for a time at an abortion clinic in the morning and a fertility clinic in the afternoon—an experience that showed him "the amazing incongruities of life." In the three decades since then, he says he has come to believe that "there's a time for fetal anesthesia, and maybe there's a time not."

IN THEIR USE OF PAIN to make the fetus seem more fully human, anti-abortion forces draw on a deep tradition. Pain has long played a special role in how society determines who is like us or not like us ("us" being those with the power to make and enforce such distinctions). The capacity to feel pain has often been put forth as proof of a common humanity. Think of Shylock's monologue in *The Merchant of Venice*: Are not Jews "hurt with the same weapons" as Christians, he demands. "If you prick us, do we not bleed?" Likewise, a presumed insensitivity to pain has been used to exclude some from humanity's privileges and protections. Many 19th-century doctors believed blacks were indifferent to pain and performed surgery on them without even that era's rudimentary anesthesia. Over time, the charmed circle of those considered alive to pain, and therefore fully human, has widened to include members of other religions and races, the poor, the criminal, the mentally ill—and, thanks to the work of Sunny Anand and others, the very young. Should the circle enlarge once more, to admit those not yet born? Should fetuses be added to what Martin Pernick, a historian of the use of anesthesia, has called "the great chain of feeling"? Anand maintains that they should.

For others, it's a harder call. When it comes to the way adults feel pain, science has borne out the optimistic belief that we are all the same under the skin. As research is now revealing, the same may not be true for fetuses; even Anand calls the fetus "a unique organism." Exhibiting his flair for the startling but apt expression, Stuart Derbyshire warns against "anthropomorphizing" the fetus, investing it with human qualities it has yet to develop. To do so, he suggests, would subtract some measure of our own humanity. And to concern ourselves only with the welfare of the fetus is to neglect the humanity of the pregnant woman, Mark Rosen notes. When considering whether to provide fetal anesthesia during an abortion, he says, it's not "erring on the safe side" to endanger a woman's health in order to prevent fetal pain that may not exist.

Indeed, the question remains just how far we would take the notion that the fetus is entitled to protection from pain. Would we be willing, for example, to supply a continuous flow of drugs to a fetus that is found to have a painful medical condition? For that matter, what about the pain of being born? Two years ago, a Swiftian satire of the Unborn Child Pain Awareness Act appeared on the progressive Web site AlterNet.org. Written by Lynn Paltrow, the executive director of the National Advocates for Pregnant Women, it urged the bill's authors to extend its provisions to those fetuses "subjected to repeated, violent maternal uterine contraction and then forced through the unimaginably narrow vaginal canal."

She continued: "Imagine the pain a fetus experiences with a forceps delivery, suffering extensive bruising during and after! Shouldn't these fetuses also be entitled to their own painkillers?" And in fact, both Nicholas Fisk and Marc Van de Velde have raised the possibility of administering pain relief to fetuses undergoing difficult deliveries. Obstetricians have yet to embrace the proposal. But Sunny Anand, for one, says the idea may have merit. Though he has "misgivings about messing with a process that has worked for thousands of years," he can envision an injection of local anesthetic into the fetus's scalp where it is grasped by the forceps or vacuum device. "Let's try and work out what's best for the baby," he says.

OLIVER SACKS

A Journey Inside the Brain

FROM *THE NEW YORK REVIEW OF BOOKS*

Reflecting on a remarkable seventy-year-old memoir of a man afflicted by a brain tumor, the neurologist Oliver Sacks finds much that is relevant to the practice of medicine today.

FRIGYES KARINTHY (BORN IN 1887 IN BUDAPEST) WAS a well-known Hungarian poet, playwright, novelist, and humorist, when he developed, at the age of forty-eight, what in retrospect were the first symptoms of a growing brain tumor.

He was having tea at his favorite café in Budapest one evening when he heard "a distinct rumbling noise, followed by a slow, increasing reverberation, . . . a louder and louder roar . . . , only to fade gradually into silence." He looked up and was surprised to see that nothing was happening. There was no train; nor, indeed, was he near a train station. "What were they playing at?" Karinthy wondered. "Trains running outside, . . . or some new means of locomotion?" It was only after the fourth "train" that he realized he was having a hallucination.

In his memoir *A Journey Round My Skull*, Karinthy reflects on how he has occasionally heard his own name whispered softly—we have all had such experiences. But this was something quite different:

> The roaring of a train [was] loud, insistent, continuous. It was powerful enough to drown real sounds. . . . After a while I realized to my astonishment that the outer world was not responsible . . . the noise must be coming from inside my head.

Many patients have described to me how they first experienced auditory hallucinations—usually not voices or noises, but music.* All of them, like Karinthy, looked around to find the source of what they were hearing, and only when they could find no possible source did they, reluctantly and sometimes fearfully, conclude that they were hallucinating. Many people in this situation fear that they are going insane—for is it not typical of madness to "hear things"?

Karinthy was not concerned on this score:

> I . . . did not find the incident at all alarming, but only very odd and unusual. . . . I could not have gone mad for, in that event, I should be incapable of diagnosing my case. Something else must be wrong. . . .

So the first chapter of his memoir (*The Invisible Train*) opens like a detective story or a mystery novel, with a puzzling and bizarre incident that reflects the changes which are starting to happen, slowly, stealthily, in his own brain. Karinthy himself would be both subject and investigator in the increasingly complex drama that he was subsequently drawn into.

* See Oliver Sacks, *Musicophilia: Tales of Music and the Brain* (Knopf, 2007); reviewed in *The New York Review of Books* by Colin McGinn, March 6, 2008.

Gifted and precocious (he had written his first novel at fifteen), Karinthy achieved fame in 1912, at the age of twenty-five, when no fewer than five of his books were published. Though he was trained in mathematics and actively interested in all aspects of science, he was especially known for his satirical writings, his political passions, and his surreal sense of humor. He had written philosophical works, plays, poems, novels, and, at the time of his first symptoms, had started writing a vast encyclopedia, which he hoped might be the twentieth-century equivalent of Diderot's monumental *Encyclopedia*. With all of this previous work, there had always been a plan, a structure, but now, forced to pay attention to what was happening in his own brain, Karinthy could only record, make notes, and reflect, without any clear notion of what lay ahead, of where this new journey would take him.

The hallucinatory train noises soon became a fixture in Karinthy's life. He started to hear them regularly, at seven o'clock each evening, whether he was in his favorite café or anywhere else. And within a few days, even stranger events started to occur:

> The mirror opposite me seemed to move. Not more than an inch or two, then it hung still. . . . But what was happening now?
>
> What was this—queer feeling—coming over me? The queerest thing was that—I didn't know what was queer. Perhaps there was nothing else queer about it. Yet I was conscious of something I had never known before, or rather I missed something I had been accustomed to since I was first conscious of being alive, though I had never paid much heed to it. I had no headache nor pain of any kind, I heard no trains, my heart was perfectly normal. And yet. . . .
>
> And yet everything, myself included, seemed to have lost its grip on reality. The tables remained in their usual places, two men were just walking across the café, and in front of me I saw the familiar water-jug and match-box. Yet in some eerie and alarming way they had all become accidental, as if they hap-

> pened to be where they were purely by chance, and might just as well be anywhere else. . . . And now the whole box of tricks was starting to roll about, as if the floor underneath it had given way. I wanted to cling on to something. . . . There wasn't a fixed point anywhere. . . . Unless, perhaps, I could find one in my own head. If I could catch hold of a single image or memory or association that would help me to recognize myself. Or even a word might do.

This is a remarkable description of what it feels like to have the very foundations of perception, of consciousness, of self, undermined—to descend (perhaps for a few moments, but they may seem an eternity) into what Proust called "the abyss of unbeing," and to long desperately for some image, some memory, some word, to pull oneself out.

At this point Karinthy started to realize that something might be seriously and strangely the matter—he wondered if he was having seizures or working up to a stroke. In the weeks that followed, he started to get further symptoms: attacks of retching and nausea, difficulties with balance and gait. He did his best to dismiss and discount these, but finally, concerned by a steadily increasing blurring of his vision, he consulted an ophthalmologist, and started on a frustrating medical odyssey:

> The doctor whom I called to consult shortly afterwards did not even examine me. Before I could describe half my symptoms he lifted his hand: "My dear fellow, you've neither aural catarrh nor have you had a stroke. . . . Nicotine poisoning, that's what's the matter with you."

Were doctors in Budapest in 1936 worse than doctors in, say, New York or London seventy years later? Not listening, not examining, being opinionated, jumping to conclusions—all are as ubiquitous, and dangerous, now as they were then and there (as Jerome Groopman describes so well in his book *How Doctors Think**). Wholly treatable disorders can go unrecognized, undiagnosed, until it is too late. Had Karinthy's first doctor examined him, he would have found a disorder of coordination indicating a cerebellar disturbance; looking into the eyes, he would have seen papilloedema—a swelling of the papillae, the optic discs—a sure sign of increased pressure in the brain. Had he paid attention to what his patient tried to tell him, he would never have been so cavalier: no one has such auditory hallucinations or sudden underminings of consciousness without a significant cerebral cause.

But Karinthy was a part of the rich and fertile café culture of Budapest, and his social circle included not only writers and artists but scientists and doctors, too. This may have made it difficult for him to get a straightforward medical opinion, for his doctors were also his friends or colleagues. As the weeks passed, Karinthy, though making light of his symptoms, started to be haunted by two memories: that of a young friend who had died of a brain tumor, and that of a film he had once seen, showing the great pioneer neurosurgeon Harvey Cushing operating on the brain of a conscious patient.

At this point, suspecting that he, too, might have a brain tumor, Karinthy insisted that the ophthalmologist, a friend of his, examine his retinas closely. His vivid recounting of this scene is both shocking and richly ironic, and shows his sharp eye and comic gifts at their best. Taken aback a little at Karinthy's insistence, the doctor who had jollied him along a few months earlier now pulled out his ophthalmoscope and looked:

> As he bent close over me, I felt the ingenious little instrument brushing my nose and I could hear him draw his breath with a slight effort, as he strained to observe me closely. I waited for the usual reassurance. "Nothing wrong there! You just want

*Houghton Mifflin, 2007.

> new glasses—a trifle stronger this time. . . ." The reality was very different. I heard Dr. H. give a sudden whistle. . . .
>
> He laid down his instrument on the table and tilted his head on one side. I saw him look at me with a kind of grave amazement, as if I had suddenly become a stranger to him.

Suddenly Karinthy ceased to be himself, a social acquaintance, an equal, a fellow human being with fears and feelings—and became a specimen. Dr. H. "was as pleasantly excited as an entomologist who has stumbled on some coveted specimen." He ran out of the room to summon his colleagues:

> In an incredibly short space of time the room was full. Assistants, house physicians, students, came pouring round, greedily snatching the ophthalmoscope from one another.
>
> The Professor himself came, turned to Dr. H., and said, "My congratulations! A really admirable diagnosis!"

As the medical men were congratulating one another, Karinthy tried to speak:

> "Gentlemen . . . !" I began modestly.
>
> Every one swung round. It was as if they had only just realized that I was of the party, and not only my papilla, which had become the centre of interest.

This scene is one that could occur, and does occur, in hospitals all over the world—the sudden focus on an intriguing pathology, and the complete forgetting of the (perhaps terrified) human being who happens to have it. All doctors are guilty of this, which is why we continue to need books from the vantage point of patients. It is salutary to be reminded by a patient as witty and observant and articulate as Karinthy of how easily the human element is apt to be forgotten in the raptures of such an "entomological" excitement.

But one needs to remember, too, how difficult and delicate an art it was, seventy years ago, to diagnose and locate a cerebral tumor. In the 1930s, there were no MRI or CT scans, only elaborate and sometimes dangerous procedures, such as injecting air into the ventricles of the brain, or injecting a dye into its blood vessels.

So it took months as Karinthy was referred from one specialist to another, and meanwhile his vision was growing worse. As he approached virtual blindness, he entered a strange world, where he could no longer be certain whether he was actually seeing or not:

> I had learnt to interpret every hint afforded by the shifting of light and to complete the general effect from memory. I was getting used to this strange semi-darkness in which I lived, and I almost began to like it. I could still see the outline of figures fairly well, and my imagination supplied the details, like a painter filling an empty frame. I tried to form a picture of any face I saw in front of me by observing the person's voice and movements. . . . The idea that I might already have gone blind struck me with sudden terror. What I fancied I saw was perhaps no more than the stuff that dreams are made on. I might only be using people's words and voices to reconstruct the lost world of reality. . . . I stood on the threshold of reality and imagination, and I began to doubt which was which. My bodily eye and my mind's eye were blending into one.

Just as Karinthy was on the verge of permanent blindness, a precise diagnosis of the tumor was made, finally, by the eminent Viennese neurologist Otto Pötzl. Immediate surgery was recommended, and Karinthy, accompanied by his wife, took a series of trains to Sweden to meet the great Herbert Olivecrona, a student of Harvey Cushing's, and one of the best neurosurgeons in the world.

Karinthy's portrait of Olivecrona, which occupies an entire chapter, is full of insight and irony, and written in a new, spare style, quite different from the lush description that precedes it. The courtesy and reserve of the cool Scandinavian neurosurgeon is delicately brought out, in contrast to the Central European emotionalism of his illustrious patient. Karinthy is done with his ambivalence, his denials, his suspicions, and he has at last found a doctor whom he can trust and even love.

He was told that the operation would last many hours, but that only local anesthetic would be used, because the brain itself has no sensory nerves, does not feel pain—and general anesthesia for such a lengthy operation was too risky. And some parts of the brain, while not sensitive to pain, may, when stimulated, evoke vivid visual or auditory memories. Karinthy described the first drilling into his skull:

> There was an infernal scream as the steel plunged into my skull. It sank more and more rapidly through the bone, and the pitch of its scream became louder and more piercing every second. . . . Suddenly, there was a violent jerk, and the noise stopped.

Karinthy heard a rush of fluid inside his head and wondered if it was blood or spinal fluid. He was then wheeled into the X-ray room, where air was injected into the ventricles of the brain to outline them and delineate the way in which they were being compressed by his tumor.

Back in the operating room, Karinthy was immobilized, face down, on the operating table, and the surgery began in earnest. The greater part of his skull was exposed, and then much of it removed, piecemeal. Karinthy felt

> a straining sensation, a feeling of pressure, a cracking sound, and a terrific wrench. . . . Something broke with a dull noise. . . . This process was repeated many times . . . like splitting open a wooden packing-case, plank by plank.

Once the skull had been opened, all pain ceased—and this itself was paradoxically disturbing:

> No, my brain did not hurt. Perhaps it was more exasperating this way than if it had. I would have preferred it to hurt me. More terrifying than any actual pain was the fact that my position seemed impossible. It was impossible for a man to be lying here with his skull open and his brain exposed to the outer world—impossible for him to lie here and live . . . impossible, incredible, indecent, for him to remain alive—and not merely alive, but conscious and in his right mind.

At intervals, the cool, kindly voice of Olivecrona broke in, explaining, reassuring, and Karinthy's apprehension was replaced by calm and curiosity. Olivecrona, here, seems almost like Virgil, guiding his poet-patient through the circles and landscapes of his brain.

Six or seven hours into the operation, Karinthy had a singular experience. It was not a dream, for he was fully conscious—though, perhaps, in an altered state of consciousness. He seemed to be looking down on his body from the ceiling of the operating theater, moving about, zooming in and out:

> The hallucination consisted in my mind seeming to move freely about the room. There was only a single light, which fell evenly on to the table. Olivecrona . . . seemed to be leaning forward . . . the lamp on his forehead threw a light into the open cavity of my skull. He had already drained off the yellowish fluid. The lobes of the cerebellum seemed to have subsided and fallen apart of themselves, and I fancied I saw the inside of the opened tumour. He had cauterized the severed veins with a red-hot electric needle. The angioma [the tumor made up of blood vessels] was already visible, lying within the cyst and a little to one side of it. The tumour itself looked like a great, red globe. In my vision it seemed as large as a small cauliflower. Its

surface was embossed so that it formed a kind of pattern, like a cameo with a carved design. . . . It seemed almost a pity that Olivecrona was to destroy it.

Karinthy's visualization or hallucination continued in minute detail. He "saw" Olivecrona skillfully removing the tumor, sucking his lower lip with concentration, and then with satisfaction that the essential part of the surgery was done.

I do not know what to call this intense visualization, informed and conjured up by his detailed knowledge of what was actually happening. Karinthy himself uses the word "hallucination," and the aerial viewpoint, looking down on one's own body, is very characteristic of what is often called an "out-of-body experience." (Such OBEs are often associated with near-death experiences such as cardiac arrest or the perception of imminent catastrophe—and they have been associated with temporal lobe seizures, or stimulation of the temporal lobes during brain surgery.)

Whatever it was, Karinthy seemed to know that the operation had been successful, that the tumor had been removed without any damage to the brain. Perhaps Olivecrona had said this to him and Karinthy had transformed his words into a vision. After this intense and reassuring vision, Karinthy fell deeply asleep and did not wake up until he was back in his own bed.

The surgery, in Olivecrona's masterly hands, had gone well—the tumor, which turned out to be benign, was gone, and Karinthy made a complete recovery, even recovering his vision, which, it had been thought, would be permanently lost. He could read and write once again, and with an exuberant sense of relief and gratitude, he rapidly composed *A Journey Round My Skull*, and sent the first copy of the German edition to the surgeon who had saved his life. He followed this with another book, *The Heavenly Report*, somewhat different in style and approach, and then started on yet another, *Message in the Bottle*. He was apparently in full health and full creative swing when he died suddenly in August of 1938. He was only fifty-one. It is said that he had a stroke while bending to tie his shoelace.

I first read *A Journey Round My Skull* as a boy of thirteen or fourteen—I think it influenced me, when I came to write my own neurological case histories—and now, rereading it sixty years later, I think it stands up remarkably well. It is not just an elaborate case history; it depicts the complex impact of a sight-, mind-, and life-threatening illness in a man of extraordinary sensibility and talent, and even something approaching genius, in the prime of his life. It becomes a journey of insight, of symbolic stages.

It has its faults: there are long digressions, philosophical and literary, where one might want a tauter narrative, and there is a certain amount of fanciful contrivance and extravagance—though this is something that Karinthy becomes more and more conscious of as he writes the book, as he is sobered by his experience, and as he tried to weld his novelistic imagination to the factual, even the clinical, realities of his situation. But despite its flaws, Karinthy's book is, to my mind, a masterpiece. We are inundated now with medical memoirs, both biographical and autobiographical—the entire genre has exploded in the last twenty years. Yet even though medical technology may have changed, the human experience has not, and *A Journey Round My Skull*, the first autobiographical description of a journey inside the brain, remains one of the very best.

DAVID WOLMAN

The Truth About Autism

FROM *WIRED*

Diagnoses of autism are on the rise, and yet it seems as though we are still learning what exactly autism is. David Wolman meets some autistic people who defy easy definitions of the disease—who subvert the idea of its being a disease at all.

THE YOUTUBE CLIP OPENS WITH A WOMAN FACING AWAY from the camera, rocking back and forth, flapping her hands awkwardly, and emitting an eerie hum. She then performs strange repetitive behaviors: slapping a piece of paper against a window, running a hand lengthwise over a computer keyboard, twisting the knob of a drawer. She bats a necklace with her hand and nuzzles her face against the pages of a book. And you find yourself thinking: Who's shooting this footage of the handicapped lady, and why do I always get sucked into watching the latest viral video?

But then the words "A Translation" appear on a black screen, and for the next five minutes, 27-year-old Amanda Baggs—who is autistic and doesn't speak—describes in vivid and articulate terms what's going on inside her head as she carries out these seemingly bizarre actions. In a synthesized voice generated by a software application, she explains that touching, tasting, and smelling allow her to have a "constant conversation" with her surroundings. These forms of nonverbal stimuli constitute her "native language," Baggs explains, and are no better or worse than spoken language. Yet her failure to speak is seen as a deficit, she says, while other people's failure to learn her language is seen as natural and acceptable.

And you find yourself thinking: She might have a point.

Baggs lives in a public housing project for the elderly and handicapped near downtown Burlington, Vermont. She has short black hair, a pointy nose, and round glasses. She usually wears a T-shirt and baggy pants, and she spends a scary amount of time—day and night—on the Internet: blogging, hanging out in Second Life, and corresponding with her autie and aspie friends. (For the uninitiated, that's *autistic* and *Asperger's*.)

On a blustery afternoon, Baggs reclines on a red futon in the apartment of her neighbor (and best friend). She has a gray travel pillow wrapped around her neck, a keyboard resting on her lap, and a DynaVox VMax computer propped against her legs.

Like many people with autism, Baggs doesn't like to look you in the eye and needs help with tasks like preparing a meal and taking a shower. In conversation she'll occasionally grunt or sigh, but she stopped speaking altogether in her early twenties. Instead, she types 120 words a minute, which the DynaVox then translates into a synthesized female voice that sounds like a deadpan British schoolteacher.

The YouTube post, she says, was a political statement, designed to call attention to people's tendency to underestimate autistics. It wasn't her first video post, but this one took off. "When the number of viewers began to climb, I got scared out of my mind," Baggs says.

As the hit count neared 100,000, her blog was flooded. At 200,000, scientists were inviting her to visit their labs. By 300,000, the TV people came calling, hearts warmed by the story of a young woman's fiery spirit and the rare glimpse into what has long been regarded as the solitary imprisonment of the autistic mind. "I've said a million times that I'm not 'trapped in my own world,'" Baggs says. "Yet what do most of these news stories lead with? Saying exactly that."

I tell her that I asked one of the world's leading authorities on autism to check out the video. The expert's opinion: Baggs must have had outside help creating it, perhaps from one of her caregivers. Her inability to talk, coupled with repetitive behaviors, lack of eye contact, and the need for assistance with everyday tasks are telltale signs of severe autism. Among all autistics, 75 percent are expected to score in the mentally retarded range on standard intelligence tests—that's an IQ of 70 or less.

People like Baggs fall at one end of an array of developmental syndromes known as autism spectrum disorders. The spectrum ranges from someone with severe disability and cognitive impairment to the socially awkward eccentric with Asperger's syndrome.

After I explain the scientist's doubts, Baggs grunts, and her mouth forms just a hint of a smirk as she lets loose a salvo on the keyboard. No one helped her shoot the video, edit it, and upload it to YouTube. She used a Sony Cybershot DSC-T1, a digital camera that can record up to 90 seconds of video (she has since upgraded). She then patched the footage together using the editing programs RAD Video Tools, VirtualDub, and DivXLand Media Subtitler. "My care provider wouldn't even know how to work the software," she says.

Baggs is part of an increasingly visible and highly networked community of autistics. Over the past decade, this group has benefited enormously from the Internet as well as innovations like type-to-speech software. Baggs may never have considered herself trapped in her own world, but thanks to technology, she can communicate with the same speed and specificity as someone using spoken language.

Autistics like Baggs are now leading a nascent civil rights movement. "I remember in '99," she says, "seeing a number of gay pride Web sites. I envied how many there were and wished there was something like that for autism. Now there is." The message: We're here. We're weird. Get used to it.

This movement is being fueled by a small but growing cadre of neuropsychological researchers who are taking a fresh look at the nature of autism itself. The condition, they say, shouldn't be thought of as a disease to be eradicated. It may be that the autistic brain is not defective but simply different—an example of the variety of human development. These researchers assert that the focus on finding a cure for autism—the disease model—has kept science from asking fundamental questions about how autistic brains function.

A cornerstone of this new approach—call it the difference model—is that past research about autistic intelligence is flawed, perhaps catastrophically so, because the instruments used to measure intelligence are bogus. "If Amanda Baggs had walked into my clinic five years ago," says Massachusetts General Hospital neuroscientist Thomas Zeffiro, one of the leading proponents of the difference model, "I would have said she was a low-functioning autistic with significant cognitive impairment. And I would have been totally wrong."

SEVENTY YEARS AGO, A BALTIMORE psychiatrist named Leo Kanner began recording observations about children in his clinic who exhibited "fascinating peculiarities." Just as Kanner's landmark paper was about to be published, a pediatrician in Vienna named Hans Asperger was putting the finishing touches on a report about a similar patient population. Both men, independently, used the same word to describe and define the condition: *autist*, or *autism*, from the Greek *autos*, meaning self.

The children had very real deficits, especially when it came to the "failure to be integrated in a social group" (Asperger) or the inborn inability to form "affective contact" with other people (Kanner). The two doctors' other observations about language impairment, repetitive behaviors, and the desire for sameness still form much of the basis of autism diagnoses in the 21st century.

On the matter of autistic intelligence, Kanner spoke of an array of mental skills, "islets of ability"—vocabulary, memory, and problem-solving that "bespeak good intelligence." Asperger, too, was struck by "a particular originality of thought and experience." Yet over the years, those islets attracted scientific interest only when they were amazing—savant-level capabilities in areas such as music, mathematics, and drawing. For the millions of people with autism who weren't savants, the general view was that their condition was tragic, their brainpower lacking.

The test typically used to substantiate this view relies heavily on language, social interaction, and cultural knowledge—areas that autistic people, by definition, find difficult. About six years ago, Meredyth Goldberg Edelson, a professor of psychology at Willamette University in Oregon, reviewed 215 articles published over the past 71 years, all making or referring to this link between autism and mental retardation. She found that most of the papers (74 percent) lacked their own research data to back up the assumption. Thirty-nine percent of the articles weren't based on any data, and even the more rigorous studies often used questionable measures of intelligence. "Are the majority of autistics mentally retarded?" Goldberg Edelson asks. "Personally, I don't think they are, but we don't have the data to answer that."

Mike Merzenich, a professor of neuroscience at UC San Francisco, says the notion that 75 percent of autistic people are mentally retarded is "incredibly wrong and destructive." He has worked with a number of autistic children, many of whom are nonverbal and would have been plunked into the low-functioning category. "We label them as retarded because they can't express what they know," and then, as they grow older, we accept that they "can't do much beyond sit in the back of a warehouse somewhere and stuff letters in envelopes."

The irony is that this dearth of data persists even as autism receives an avalanche of attention. Organizations such as Autism Speaks advocate for research and resources. Celebrity parents like Toni Braxton, Ed Asner, and Jenny McCarthy get high-profile coverage on talk shows and TV news-magazines. Newsweeklies raise fears of an autism epidemic. But is there an epidemic? There's certainly the perception of one. According to the Centers for Disease Control, one out of every 150 8-year-old children (in the areas of the US most recently studied) has an autism spectrum disorder, a prevalence much higher than in decades past, when the rate was thought to be in the range of four or five cases per 10,000 children. But no one knows whether this apparent explosion of cases is due to an actual rise in autism, changing diagnostic criteria, inconsistent survey techniques, or some combination of the three.

In his original paper in 1943, Kanner wrote that while many of the children he examined "were at one time or another looked upon as feebleminded, they are all unquestionably endowed with *good cognitive potentialities.*" Sixty-five years later, though, little is known about those potentialities. As one researcher told me, "There's no money in the field for looking at differences" in the autistic brain. "But if you talk about trying to fix a problem—then the funding comes."

On the outskirts of Montreal sits a brick monolith, the Hôpital Rivière-des-Prairies. Once one of Canada's most notorious asylums, it now has a small number of resident psychiatric patients, but most of the space has been converted into clinics and research facilities.

One of the leading researchers here is Laurent Mottron, 55, a psychiatrist specializing in autism. Mottron, who grew up in postwar France, had a tough childhood. His family had a history of schizophrenia and Tourette syndrome, and he probably has what today would be diagnosed as attention deficit and hyperactivity disorder. Naturally, he went into psychiatry. By the early '80s, Mottron was doing clinical work at a school in Tours that catered to children with sensory impairment, including autism. "The view then," Mottron says, "was that these children could be reeled back to normalcy with play therapy and work on the parents' relationships"—a gentle way of saying that the parents, especially the mother, were to blame. (The theory that emotionally distant "refrigerator mothers" caused autism had by then been rejected in the US, but in France and many other countries, the view lingered.)

After only a few weeks on the job, Mottron decided the theories were crap. "These children were just of another kind," he says. "You couldn't turn someone autistic or make someone not autistic. It was hardwired." In 1986, Mottron began working with an autistic man who would later become known in the scientific literature as "E.C." A draftsman who specialized in mechanical drawings, E.C. had incredible savant skills in 3-D drawing. He could rotate objects in his mind and make technical drawings without the need for a single revision. After two years of working with E.C., Mottron made his second breakthrough—not about autistics this time but about the rest of us: People with standard-issue brains—so-called neurotypicals—don't have the perceptual abilities to do what E.C. could do. "It's just inconsistent with how our brains work," Mottron says.

From that day forward, he decided to challenge the disease model underlying most autism research. "I wanted to go as far as I could to show that their perceptions—their brains—are totally different." Not damaged. Not dysfunctional. Just different.

By the mid-1990s, Mottron was a faculty member at the University of Montreal, where he began publishing papers on "atypicalities of perception" in autistic subjects. When performing certain mental tasks—especially when tapping visual, spatial, and auditory functions—autistics have shown superior performance compared with neurotypicals. Call it the upside of autism. Dozens of studies—Mottron's and others—have demonstrated that people with autism spectrum disorder have a number of strengths: a higher prevalence of perfect pitch, enhanced ability with 3-D drawing and pattern recognition, more accurate graphic recall, and various superior memory skills.

Yet most scientists who come across these skills classify them as "anomalous peaks of ability," set them aside, and return to the questions that drive most research: What's wrong with the autistic brain? Can we find the genes responsible so that we can someday cure it? Is there a unifying theory of autism? With severe autistics, cognitive strengths are even more apt to be overlooked because these individuals have such obvious deficits and are so hard to test. People like Baggs don't speak, others may run out of the room, and still others might not be able to hold a pencil. And besides, if 75 percent of them are mentally retarded, well, why bother?

Mottron draws a parallel with homosexuality. Until 1974, psychiatry's bible, the *Diagnostic and Statistical Manual of Mental Disorders*, described being gay as a mental illness. Someday, Mottron says, we'll look back on today's ideas about autism with the same sense of shame that we now feel when talking about psychology's pre-1974 views on sexuality. "We want to break the idea that autism should definitely be suppressed," he says.

MICHELLE DAWSON DOESN'T DRIVE OR cook. Public transit overwhelms her, and face-to-face interaction is an ordeal. She was employed as a postal worker in 1998 when she "came out of the closet" with her diagnosis of autism, which she received in the early '90s. After that, she claims, Canada Post harassed her to such a degree that she was forced to take a permanent leave of absence, starting in 2002. (Canada Post says Dawson was treated fairly.) To fight back, she went on an information-devouring rampage. "There's such a variety of human behavior. Why is my kind wrong?" she asks. She eventually began scouring the libraries of McGill University in Montreal to delve into the autism literature. She searched out journal articles using the online catalog and sat on the floor reading studies among the stacks.

Dawson, like Baggs, has become a reluctant spokesperson for this new view of autism. Both are prolific bloggers and correspond constantly with scientists, parents' groups, medical institutions, the courts, journalists, and anyone else who'll listen to their stories of how autistics are mistreated. Baggs has been using YouTube to make her point; Dawson's weapon is science.

In 2001, Dawson contacted Mottron, figuring that his clinic might help improve the quality of her life. Mottron tried to give her some advice on navigating the neurotypical world, but his tips on how to handle banking, shopping, and buses didn't help. After meeting with her a few times, Mottron began to suspect that what Dawson really needed was a sense of purpose. In 2003, he handed her one of his in-progress journal articles and asked her to copy-edit the grammar. So Dawson started reading. "I criticized his science almost immediately," she says.

Encouraged by Dawson's interest, Mottron sent her other papers. She responded with written critiques of his work. Then one day in early 2003, she called with a question. "I asked: 'How did they control for attention in that fMRI face study?' That caught his attention." Dawson had flagged an error that Mottron says most postdocs would have missed. He was impressed, and over the next few months he sought Dawson's input on other technical questions. Eventually, he invited her to collaborate with his research group, despite the fact that her only academic credential was a high school diploma.

Dawson has an incredible memory, but she's not a savant. What makes her unique, Mottron says, is her gift for scientific analysis—the way she can sniff through methodologies and statistical manipulation, hunting down tiny errors and weak links in logic.

Last summer, the peer-reviewed journal *Psychological Science* published a study titled "The Level and Nature of Autistic Intelligence." The lead author was Michelle Dawson. The paper argues that autistic smarts have been underestimated because the tools for assessing intelligence depend on techniques ill-suited to autistics. The researchers administered two different intelligence tests to 51 children and adults diagnosed with autism and to 43 non-autistic children and adults.

The first test, known as the Wechsler Intelligence Scale, has helped solidify the notion of peaks of ability amid otherwise pervasive mental retardation among autistics. The other test is Raven's Progressive Matrices, which requires neither a race against the clock nor a proctor breathing down your neck. The Raven is considered as reliable as the Wechsler, but the Wechsler is far more commonly used. Perhaps that's because it requires less effort for the average test taker. Raven measures abstract reasoning—"effortful" operations like spotting patterns or solving geometric puzzles. In contrast, much of the Wechsler assesses crystallized skills like acquired vocabulary, making correct change, or knowing that milk goes in the fridge and cereal in the cupboard—learned information that most people intuit or recall almost automatically.

What the researchers found was that while non-autistic subjects scored just about the same—a little above average—on both tests, the autistic group scored much better on the Raven. Two individuals' scores swung from the mentally retarded range to the 94th percentile. More significantly, the subset of autistic children in the study scored roughly 30 percentile points higher on the Raven than they did on the more language-dependent Wechsler, pulling all but a couple of them out of the range for mental retardation.

A number of scientists shrugged off the results—of course autistics would do better on nonverbal tests. But Dawson and her coauthors saw something more. The "peaks of ability" on the Wechsler correlated strongly with the average scores on the Raven. The finding suggests the Wechsler scores give only a glimpse of the autistics' intelligence, whereas the Raven—the gold standard of fluid intelligence testing—reveals the true, or at least truer, level of general intelligence.

Yet to a remarkable degree, scientists conducting cognitive evaluations continue to use tests which presume that people who can't communicate the answer don't know the answer. The question is:

Why? Greg Allen, an assistant professor of psychiatry at University of Texas Southwestern Medical Center, says that although most researchers know the Wechsler doesn't provide a good assessment of people with autism, there's pressure to use the test anyway. "Say you're submitting a grant to study autistic people by comparing them to a control group," he says. "The first question that comes up is: Did you control for IQ? Matching people on IQ is meant to clean up the methodology, but I think it can also end up damaging the study."

And that hurts autistic people, Dawson says. She makes a comparison with blindness. Of course blind people have a disability and need special accommodation. But you wouldn't give a blind person a test heavily dependent on vision and interpret their poor score as an accurate measure of intelligence. Mottron is unequivocal: Because of recent research, especially the Raven paper, it's clearer than ever that so-called low-functioning people like Amanda Baggs are more intelligent than once presumed. The Dawson paper was hardly conclusive, but it generated buzz among scientists and the media. Mottron's team is now collaborating with Massachusetts General Hospital's Zeffiro, a neuroimaging expert, to dig deeper. Zeffiro and company are looking for variable types of mental processing *without* asking, what's wrong with this brain? Their first study compares fMRI results from autistic and control subjects whose brains were imaged while they performed the Raven test. The group is currently crunching numbers for publication, and the study looks both perplexing and promising.

Surprisingly, they didn't find any variability in which parts of the brain lit up when subjects performed the tasks. "We thought we'd see different patterns of activation," Zeffiro says, "but it looks like the similarities outweigh the dissimilarities." When they examined participants' Raven scores together with response times, however, they noticed something odd. The two groups had the same error rates, but as an aggregate, the autistics completed the tasks 40 percent faster than the non-autistics. "They spent less time coming up with the same number of right answers. The only explanation we can see right now," Zeffiro says, is that autistic brains working on this set of tasks "seem to be engaged at a higher level of efficiency." That may have to do with greater connectivity within an area or areas of the brain. He and other researchers are already exploring this hypothesis using diffusion tensor imaging, which measures the density of brain wiring.

But critics of the difference model reject the whole idea that autism is merely another example of neuro-diversity. After all, being able to plan your meals for the week or ask for directions bespeaks important forms of intelligence. "If you pretend the areas that are troubled aren't there, you miss important aspects of the person," says Fred Volkmar, director of Yale's Child Study Center.

In the vast majority of journal articles, autism is referred to as a disorder, and the majority of neuro-psychiatric experts will tell you that the description fits—something is wrong with the autistic brain. UCSF's Merzenich, who agrees that conventional intelligence-testing tools are misleading, still doesn't think the difference model makes sense. Many autistics are probably smarter than we think, he says. But there's little question that more severe autism is characterized by what Merzenich terms "grossly abnormal" brain development that can lead to a "catastrophic end state." Denying this reality, he says, is misguided. Yale's Volkmar likens it to telling a physically disabled person: "You don't need a wheelchair. Walk!"

Meanwhile parents, educators, and autism advocates worry that focusing on the latent abilities and intelligence of autistic people may eventually lead to cuts in funding both for research into a cure and services provided by government. As one mother of an autistic boy told me, "There's no question that my son needs treatment and a cure."

BACK IN BURLINGTON, BAGGS IS cueing up another YouTube clip. She angles her computer screen so I can see it. Set to the soundtrack of Queen's "Under Pressure," it's a montage of close-up videos showing behaviors like pen clicking, thumb twiddling, and finger tapping. The message: Why are some stress-related behaviors socially permissible, while others—like the rocking bodies and flapping arms commonly associated with autism—are not? Hit count for the video at last check: 80,000 and climbing.

Should autism be treated? Yes, says Baggs, it should be treated with respect. "People aren't interested in us functioning with the brains we have," she says, because autism is considered to be outside the range of normal variability. "I don't fit the stereotype of autism. But who does?" she asks, hammering especially hard on the keyboard. "The definition of autism is so fluid and changing every few years." What's exciting, she says, is that Mottron and other scientists have "found universal strengths where others usually look for universal deficits." Neuro-cognitive science, she says, is finally catching up to what she and many other adults with autism have been saying all along.

Baggs is working on some new videos. One project is tentatively titled "Am I a Person Yet?" She'll explore communication, empathy, self-reflection—core elements of the human experience that have at times been used to define personhood itself. And at various points during the clip, she'll ask: "Am I a person yet?" It's a provocative idea, and you might find yourself thinking: She has a point.

ALEX KOTLOWITZ

Blocking the Transmission of Violence

FROM *THE NEW YORK TIMES MAGAZINE*

Powerful, sustained and early intervention can be, as doctors know, an effective way to halt the spread of a viral outbreak. Can the same approach work to stop violence? Alex Kotlowitz profiles a group of "interrupters," many of them ex-cons, who, led by an epidemiologist, are putting this idea into practice on the streets of Chicago.

LAST SUMMER, MARTIN TORRES WAS WORKING AS A COOK in Austin, Tex., when, on the morning of Aug. 23, he received a call from a relative. His 17-year-old nephew, Emilio, had been murdered. According to the police, Emilio was walking down a street on Chicago's South Side when someone shot him in the chest, possibly the culmination of an ongoing dispute. Like many killings, Emilio's received just a few sentences in the local newspapers. Torres, who was especially close to his nephew, got on the first Greyhound bus to Chicago. He was grieving and plotting retribution. "I thought, Man, I'm going to take care of business," he told me recently. "That's how I live. I was going hunting. This is my own blood, my nephew."

Torres, who is 38, grew up in a dicey section of Chicago, and even by the standards of his neighborhood he was a rough character. His nickname was Packman, because he was known to always pack a gun. He was first shot when he was 12, in the legs with buckshot by members of a rival gang. He was shot five more times, including once through the jaw, another time in his right shoulder and the last time—seven years ago—in his right thigh, with a .38-caliber bullet that is still lodged there. On his chest, he has tattooed a tombstone with the name "Buff" at its center, a tribute to a friend who was killed on his 18th birthday. Torres was the head of a small Hispanic gang, and though he is no longer active, he still wears two silver studs in his left ear, a sign of his affiliation.

When he arrived in Chicago, he began to ask around, and within a day believed he had figured out who killed his nephew. He also began drinking a lot—mostly Hennessey cognac. He borrowed two guns, a .38 and a .380, from guys he knew. He would, he thought, wait until after the funeral to track down his nephew's assailants.

Zale Hoddenbach looks like an ex-military man. He wears his hair cropped and has a trimmed goatee that highlights his angular jaw. He often wears T-shirts that fit tightly around his muscled arms, though he also carries a slight paunch. When he was younger, Hoddenbach, who is also 38, belonged to a gang that was under the same umbrella as Torres's, and so when the two men first met 17 years ago at Pontiac Correctional Center, an Illinois maximum-security prison, they became friendly. Hoddenbach was serving time for armed violence; Torres for possession of a stolen car and a gun (he was, he says, on his way to make a hit). "Zale was always in segregation, in the hole for fights," Torres told me. "He was aggressive." In one scuffle, Hoddenbach lost the sight in his right eye after an inmate pierced it with a shank. Torres and Hoddenbach were at Pontiac together for about a year but quickly lost touch after they were both released.

Shortly after Torres arrived in Chicago last summer, Hoddenbach received a phone call from Torres's brother, the father of the young man who was murdered. He was worried that Torres was preparing to seek revenge and hoped that Hoddenbach would speak with him. When Hoddenbach called, Torres was thrilled. He immediately thought that his old prison buddy was going to join him in his search for the killer. But instead Hoddenbach tried to talk him down, telling him retribution wasn't what his brother wanted. "I didn't understand what the hell he was talking about," Torres told me when I talked to him six months later. "This didn't seem like the person I knew." The next day Hoddenbach appeared at the wake, which was held at New Life Community Church, housed in a low-slung former factory. He spent the day by Torres's side, sitting with him, talking to him, urging him to respect his brother's wishes. When Torres went to the parking lot for a smoke, his hands shaking from agitation, Hoddenbach would follow. "Because of our relationship, I thought there was a chance," Hoddenbach told me. "We were both cut from the same cloth." Hoddenbach knew from experience that the longer he could delay Torres from heading out, the more chance he'd have of keeping him from shooting someone. So he let him vent for a few hours. Then Hoddenbach started laying into him with every argument he could think of: *Look around, do you see any old guys here? I never seen so many young kids at a funeral. Look at these kids, what does the future hold for them? Where do we fit in? Who are you to step on your brother's wishes?*

THE STUBBORN CORE OF VIOLENCE in American cities is troubling and perplexing. Even as homicide rates have declined across the country—in some places, like New York, by a remarkable amount—gunplay continues to plague economically struggling minority communities. For 25 years, murder has been the leading cause of death among African-American men between the ages of 15 and 34, according to the Centers for Disease Control and Prevention, which has analyzed data up to 2005. And the past few years have seen an uptick in homicides in many cities. Since 2004, for instance, they are up 19 percent in Philadelphia and Milwaukee, 29 percent in Houston and 54 percent in Oakland. Just two weekends ago in Chicago, with the first warm weather, 36 people were shot, 7 of them fatally. The *Chicago Sun-Times* called it the "weekend of rage." Many killings are attributed to gang conflicts and are confined to particular neighborhoods. In Chicago, where on average five people were shot each day last year, 83 percent of the assaults were concentrated in half the police districts. So for people living outside those neighborhoods, the frequent outbursts of unrestrained anger have been easy to ignore. But each shooting, each murder, leaves a devastating legacy, and a growing school of thought suggests that there's little we can do about the entrenched urban poverty if the relentless pattern of street violence isn't somehow broken.

The traditional response has been more focused policing and longer prison sentences, but law enforcement does little to disrupt a street code that allows, if not encourages, the settling of squabbles with deadly force. Zale Hoddenbach, who works for an organization called CeaseFire, is part of an unusual effort to apply the principles of public health to the brutality of the streets. CeaseFire tries to deal with these quarrels on the front end. Hoddenbach's job is to suss out smoldering disputes and to intervene before matters get out of hand. His job title is violence interrupter, a term that while not artful seems bluntly self-explanatory. Newspaper accounts usually refer to the organization as a gang-intervention program, and Hoddenbach and most of his colleagues are indeed former gang leaders. But CeaseFire doesn't necessarily aim to get people out of gangs—nor interrupt the drug trade. It's almost blindly focused on one thing: preventing shootings.

CeaseFire's founder, Gary Slutkin, is an epidemiologist and a physician who for 10 years battled infectious diseases in Africa. He says that violence directly mimics infections like tuberculosis and AIDS, and so, he suggests, the treatment ought to mimic the regimen applied to these diseases: go after the most infected, and stop the infection at its source. "For violence, we're trying to interrupt the next event, the next transmission, the next violent activity," Slutkin told me recently. "And the violent activity predicts the next violent activity like H.I.V. predicts the next H.I.V. and TB predicts the next TB." Slutkin wants to shift how we think about violence from a moral issue (good and bad people) to a public health one (healthful and unhealthful behavior).

EVERY WEDNESDAY AFTERNOON, IN A Spartan room on the 10th floor of the University of Illinois at Chicago's public-health building, 15 to 25 men—and two women—all violence interrupters, sit around tables arranged in a circle and ruminate on the rage percolating in the city. Most are in their 40s and 50s, though some, like Hoddenbach, are a bit younger. All of them are black or Hispanic and in one manner or another have themselves been privy to, if not participants in, the brutality of the streets.

On a Wednesday near the end of March, Slutkin made a rare appearance; he ordinarily leaves the day-to-day operations to a staff member. Fit at 57, Slutkin has a somewhat disheveled appearance—tie askew, hair uncombed, seemingly forgetful. Some see his presentation as a calculated effort to disarm. "Slutkin does his thing in his Slutkinesque way," notes Carl Bell, a psychiatrist who has long worked with children exposed to neighborhood violence and who admires CeaseFire's work. "He seems kind of disorganized, but he's not." Hoddenbach told me: "You can't make too much of that guy. In the beginning, he gives you that look like he doesn't know what you're talking about."

Slutkin had come to talk with the group about a recent high-profile incident outside Crane Tech High School on the city's West Side. An 18-year-old boy was shot and died on the school's steps, while nearby another boy was savagely beaten with a golf club. Since the beginning of the school year, 18 Chicago public-school students had been killed. (Another six would be murdered in the coming weeks.) The interrupters told Slutkin that there was a large police presence at the school, at least temporarily muffling any hostilities there, and that the police were even escorting some kids to and from school. They then told him what was happening off the radar in their neighborhoods. There was the continuing discord at another high school involving a group of girls ("They'd argue with a stop sign," one of the interrupters noted); a 14-year-old boy with a gang tattoo on his forehead was shot by an older gang member just out of prison; a 15-year-old was shot in the stomach by a rival gang member as he came out of his house; and a former CeaseFire colleague was struggling to keep himself from losing control after his own sons were beaten. There was also a high-school basketball player shot four times; a 12-year-old boy shot at a party; gang members arming themselves to counter an egging of their freshly painted cars; and a high-ranking gang member who was on life support after being shot, and whose sister was overheard talking on her cellphone in the hospital, urging someone to "get those straps together. Get loaded."

These incidents all occurred over the previous seven days. In each of them, the interrupters had stepped in to try to keep one act of enmity from spiraling into another. Some had more success than others. Janell Sails prodded the guys with the egged cars to go to a car wash and then persuaded them it wasn't worth risking their lives over a stupid prank. At Crane Tech High School, three of the interrupters fanned out, trying to convince the five gangs involved in the conflict to lie low, but they conceded that they were unable to reach some of the main players. Many of the interrupters seem bewildered by what they see as a wilder group of youngsters now running the streets and by a gang structure that is no longer top-down but is instead made up of many small groups—which they refer to as cliques—whose members are answerable to a handful of peers.

For an hour, Slutkin leaned on the table, playing with a piece of

Scotch tape, keenly listening. In some situations, Slutkin can appear detached and didactic. He can wear people down with his long discourses, and some of the interrupters say they sometimes tune him out. (On one occasion, he tried to explain to me the relationship between emotional intelligence and quantum physics.) But having seen a lot of out-of-control behavior, Slutkin is a big believer in controlling emotions. So he has taught himself not to break into discussions and to digest before presenting his view. The interrupters say he has their unqualified loyalty. Hoddenbach told me that he now considers Slutkin a friend.

It became clear as they delivered their reports that many of the interrupters were worn down. One of them, Calvin Buchanan, whose street name is Monster and who just recently joined CeaseFire, showed the others six stitches over his left eye; someone had cracked a beer bottle on his head while he was mediating an argument between two men. The other interrupters applauded when Buchanan told them that, though tempted, he restrained himself from getting even.

When Slutkin finally spoke, he first praised the interrupters for their work. "Everybody's overreacting, and you're trying to cool them down," he told them. He then asked if any of them had been experiencing jitteriness or fear. He spent the next half-hour teaching stress-reduction exercises. If they could calm themselves, he seemed to be saying, they could also calm others. I recalled what one of the interrupters told me a few weeks earlier: "We helped create the madness, and now we're trying to debug it."

IN THE PUBLIC-HEALTH FIELD, THERE have long been two schools of thought on derailing violence. One focuses on environmental factors, specifically trying to limit gun purchases and making guns safer. The other tries to influence behavior by introducing school-based curricula like antidrug and safe-sex campaigns.

Slutkin is going after it in a third way—as if he were trying to contain an infectious disease. The fact that there's no vaccine or medical cure for violence doesn't dissuade him. He points out that in the early days of AIDS, there was no treatment either. In the short run, he's just trying to halt the spread of violence. In the long run, though, he says he hopes to alter behavior and what's considered socially acceptable.

Slutkin's perspective grew out of his own experience as an infectious-disease doctor. In 1981, six years out of the University of Chicago Pritzker School of Medicine, Slutkin was asked to lead the TB program in San Francisco. With an influx of new refugees from Cambodia, Laos and Vietnam, the number of cases in the city had nearly doubled. Slutkin chose to concentrate on those who had the most active TB; on average, they were infecting 6 to 10 others a year. Slutkin hired Southeast Asian outreach workers who could not only locate the infected individuals but who could also stick with them for nine months, making sure they took the necessary medication. These outreach workers knew the communities and spoke the languages, and they were able to persuade family members of infected people to be tested. Slutkin also went after the toughest cases—26 people with drug-resistant TB. The chance of curing those people was slim, but Slutkin reckoned that if they went untreated, the disease would continue to spread. "Gary wasn't constrained by the textbook," says Eric Goosby, who worked in the clinic and is now the chief executive of the Pangaea Global AIDS Foundation. Within two years, the number of TB cases, at least among these new immigrants, declined sharply.

Slutkin then spent 10 years in Africa, first in refugee camps in Somalia and then working, in Uganda and other countries, for the World Health Organization to curtail the spread of AIDS. During his first posting, in Somalia, a cholera epidemic spread from camp to camp. Slutkin had never dealt with an outbreak of this sort, and he was overwhelmed. The diarrhea from cholera is so severe that patients can die within hours from dehydration. According to Sandy Gove, who was then married to Slutkin and was also a doctor in the camps, infection rates were approaching 10 percent; in one camp there were 1,000 severely ill refugees. "It was desperate," she told me. Slutkin drove a Land Cruiser two and a half days to an American military base along the coast to the closest phone. He called doctors in Europe and the United States, trying to get information. He also asked the soldiers at the base for blue food coloring, which he then poured into the water sources of the bacteria, a warning to refugees not to drink. "What Gary is really good about is laying out a broad strategic plan and keeping ahead of something," Gove told me. There were only six doctors for the 40 refugee camps, so Slutkin and Gove trained birth attendants to spot infected people and to give them rehydration therapies in their homes. Because the birth attendants were refugees, they were trusted and could persuade those with the most severe symptoms to receive aid at the medical tent.

After leaving Africa, Slutkin returned to Chicago, where he was raised and where he could attend to his aging parents. (He later remarried there.) It was 1995, and there had been a series of horrific murders involving children in the city. He was convinced that longer sentences and more police officers had made little difference. "Punishment doesn't drive behavior," he told me. "Copying and modeling and the social expectations of your peers is what drives your behavior."

Borrowing some ideas (and the name) from a successful Boston program, Slutkin initially established an approach that exists in one form or another in many cities: outreach workers tried to get youth and young adults into school or to help them find jobs. These outreach workers were also doing dispute mediation. But Slutkin was feeling his way, much as he had in Somalia during the cholera epidemic. One of Slutkin's colleagues, Tio Hardiman, brought up an uncomfortable truth: the program wasn't reaching the most bellicose, those most likely to pull a trigger. So in 2004, Hardiman suggested that, in addition to outreach workers, they also hire men and women who had been deep into street life, and he began recruiting people even while they were still in prison. Hardiman told me he was looking for those "right there on the edge." (The interrupters are paid roughly $15 an hour, and those working full time receive benefits from the University of Illinois at Chicago, where CeaseFire is housed.) The new recruits, with strong connections to the toughest communities, would focus solely on sniffing out clashes that had the potential to escalate. They would intervene in potential acts of retribution—as well as try to defuse seemingly minor spats that might erupt into something bigger, like disputes over women or insulting remarks.

As CeaseFire evolved, Slutkin says he started to realize how much it was drawing on his experiences fighting TB and AIDS. "Early intervention in TB is actually treatment of the most infectious people," Slutkin told me recently. "They're the ones who are infecting others. So treatment of the most infectious spreaders is the most effective strategy known and now accepted in the world." And, he continued, you want to go after them with individuals who themselves were once either infectious spreaders or at high risk for the illness. In the case of violence, you use those who were once hard-core, once the most belligerent, once the most uncontrollable, once the angriest. They are the most convincing messengers. It's why, for instance, Slutkin and his colleagues asked sex workers in Uganda and other nations to spread the word to other sex workers about safer sexual behavior. Then, Slutkin said, you train them, as you would paraprofessionals, as he and Gove did when they trained birth attendants to spot cholera in Somalia.

The first step to containing the spread of an infectious disease is minimizing transmission. The parallel in Slutkin's Chicago work is thwarting retaliations, which is precisely what Hoddenbach was trying to do in the aftermath of Emilio Torres's murder. But Slutkin is also looking for the equivalent of a cure. The way public-health doctors think of curing disease when there are no drug treatments is by changing behavior. Smoking is the most obvious example. Cigarettes are still around. And there's no easy remedy for lung cancer or emphysema. So the best way to deal with the diseases associated with smoking is to get people to stop smoking. In Uganda, Slutkin and his colleagues tried to change behavior by encouraging people to have fewer sexual partners and to use condoms. CeaseFire has a visible public-communications campaign, which includes billboards and bumper stickers (which read, "Stop. Killing. People"). It also holds rallies—or what it calls "responses"—at the sites of killings. But much research suggests that peer or social pressure is the most effective way to change behavior. "It was a real turning point for me," Slutkin said, "when I was working on the AIDS epidemic and saw research findings that showed that the principal determinant of whether someone uses a condom or not is whether they think their friends use them." Daniel Webster, a professor of public health at Johns Hopkins University who has looked closely at CeaseFire, told me, "The guys out there doing the interruption have some prestige and reputation, and I think the hope is that they start to change a culture so that you can retain your status, retain your manliness and be able to walk away from events where all expectations were that you were supposed to respond with lethal force."

As a result, the interrupters operate in a netherworld between upholding the law and upholding the logic of the streets. They're not meant to be a substitute for the police, and indeed, sometimes the interrupters negotiate disputes involving illicit goings-on. They often walk a fine line between mediating and seeming to condone criminal activity. At one Wednesday meeting this past December, the interrupters argued over whether they could dissuade stickup artists from shooting their victims; persuading them to stop robbing people didn't come up in the discussion.

LAST DECEMBER, AT THE FIRST Wednesday meeting I attended, James Highsmith came up to introduce himself. At 58, Highsmith is one of the older interrupters. He wears striped, collared shirts, black rectangular glasses and often a black Borsalino, an Italian-made fedora. He reminded me that I had mentioned him in my book, *There Are No Children Here*, about life in a Chicago public-housing project in the late 1980s. I wrote about a picnic that some Chicago drug kingpins gave in a South Side park. There was a car show, a wet T-shirt contest and softball games for the children. About 2,000 people attended, dancing to a live band while the drug lords showed off their Mercedes Benzes, Rolls-Royces and Jaguars. Highsmith was the key sponsor of the event. He controlled the drug trade on the city's South Side. He owned a towing business, an auto-mechanic's shop and a nightclub, as well as a 38-foot boat. In January 1994, he was sentenced to 14 years in federal prison on drug-conspiracy charges; he was released in 2004. Highsmith was just the kind of recruit CeaseFire looks for: an older man getting out of the penitentiary who once had standing on the streets and who, through word of mouth, appears ready, eager even, to discard his former persona. "I'm a work in progress," Highsmith told me.

One evening we were sitting in Highsmith's basement apartment when the phone rang. It was Alphonso Prater, another interrupter. The two had a reunion of sorts when they joined CeaseFire; they shared a cell in the county jail 34 years ago. Prater's voice is so raspy it sounds as if he has gravel in his throat. He told me that he became permanently hoarse after a long stint in segregation in prison; he had to shout to talk with others. When Prater called the night I was there, all Highsmith could make out was: "There's some high-tech stuff going on. I need you to talk to some folks." Highsmith didn't ask any questions.

We drove to a poorly lighted side street on the city's West Side. Empty beer bottles littered the side of the road. Prater, who is short and wiry and has trouble keeping still, was bouncing on the sidewalk, standing next to a lanky middle-aged man who had receded into his oversize hooded sweatshirt. Highsmith, Prater and another interrupter joined the man in a parked car, where they talked for half an hour. When they were done, the car peeled away, two other sedans escorting it, one in front, the other in the rear. "Protection," Highsmith commented. Apparently, the man in the hooded sweatshirt, whom I would meet later, had been an intermediary in a drug deal. He had taken an out-of-town buyer holding $30,000 in cash to a house on the South Side to buy drugs. But when they got there, they were met by six men in the backyard, each armed with a pistol or an automatic weapon, and robbed. The out-of-town buyer believed he'd been set up by the intermediary, who, in turn, was trying to hunt down the stickup artists. In the car, Prater, who knew the intermediary, had worked to cool him down, while Highsmith promised to see if he could find someone who might know the stickup guys and could negotiate with them. The intermediary told Prater and Highsmith, a bit ominously, "Something got to give."

After the intermediary drove off, Prater joked that there was no way he was getting back in a car with him, that he was too overheated and too likely to be the target or the shooter. "I'm not sure we can do anything about this one," Highsmith told Prater.

RELYING ON HARDENED TYPES—THE ONES who, as Webster of Johns Hopkins says, have some prestige on the streets—is risky. They have prestige for a reason. Hoddenbach, who once beat someone so badly he punctured his lungs, is reluctant to talk about his past. "I don't want to be seen as a monster," he told me. Hoddenbach's ethnicity is hard to pinpoint. His father was Dutch and his mother Puerto Rican, and he's so light-skinned his street name was Casper. He has a discerning gaze and mischievous smile, and can be hardheaded and impatient. (At the Wednesday meetings, he often sits near the door and whispers entreaties to the others to speed things up.) Hoddenbach's father had an explosive temper, and to steal from Slutkin's lingo, he seems to have infected others. Two of Hoddenbach's older brothers are serving time for murder. His third brother has carved out a legitimate life as a manager at a manufacturing firm. Hoddenbach always worked. He did maintenance on train equipment and towed airplanes at a private airport. But he was also active in a Hispanic street gang and was known for his unmitigated aggression. He served a total of eight years in the state penitentiary, the last stay for charges that included aggravated battery. He was released in 2002.

In January, I was with Slutkin in Baltimore, where he spoke about CeaseFire to a small gathering of local civic leaders at a private home. During the two-hour meeting, Slutkin never mentioned that the interrupters were ex-felons. When I later asked him about that omission, he conceded that talking about their personal histories "is a dilemma. I haven't solved it." I spent many hours with Hoddenbach and the others, trying to understand how they chose to make the transition from gangster to peacemaker, how they put thuggery behind them. It is, of course, their street savvy and reputations that make them effective for CeaseFire. (One supporter of the program admiringly called it "a terrifying strategy" because of the inherent risks.) Some CeaseFire workers have, indeed, reverted to their old ways. One outreach worker was fired after he was arrested for possession of an AK-47 and a handgun. Another outreach worker and an interrupter were let go after they were arrested for dealing drugs. Word-of-mouth allegations often circulate, and privately, some in the police department worry about CeaseFire's workers returning to their old habits.

Not all the interrupters I talked to could articulate how they had made the transition. Some, like Hoddenbach, find religion—in his case, Christianity. He also has four children he feels responsible for, and has found ways to decompress, like going for long runs. (His brother Mark speculated that "maybe he just wants to give back what he took out.") I once asked Hoddenbach if he has ever apologized to anyone he hurt. We were with one of his old friends from the street, who started guffawing, as if I had asked Hoddenbach if he ever wore dresses. "I done it twice," Hoddenbach told us—quickly silencing his friend and saving me from further embarrassment. (One apology was to the brother of the man whose lungs he'd punctured; the other was to a rival gang member he shot.) Alphonso Prater told me that the last time he was released from prison, in 2001, an older woman hired him to gut some homes she was renovating. She trusted him with the keys to the homes, and something about that small gesture lifted him. "She seen something in me that I didn't see," he told me.

Though the interrupters may not put it this way, the Wednesday meetings are a kind of therapy. One staff member laughingly compared it to a 12-step program. It was clear to me that they leaned on one another—a lot. Prater once got an urgent call from his daughter, who said her boyfriend was beating her. Prater got in his car and began to race to her house; as he was about to run a stop sign, he glimpsed a police car on the corner. He skidded to a halt. It gave him a moment to think, and he called his CeaseFire supervisor, Tio Hardiman, who got another interrupter to visit Prater's daughter. Not long ago, three old-timers fresh out of prison ruthlessly ridiculed Hoddenbach for his work with CeaseFire. They were relentless, and Hoddenbach asked to sit down with them. But when it came time to meet, he realized he was too riled, and so he asked another interrupter, Tim White, to go in his place. "I was worried I was going to whip their asses, and wherever it went from there it went," Hoddenbach told me. "They were old feelings, feelings I don't want to revisit."

Recently I went out to lunch with Hoddenbach and Torres. It had been four months since Torres buried his nephew. Torres, who looked worn and agitated (he would get up periodically to smoke a cigarette outside), seemed paradoxically both grateful to and annoyed at Hoddenbach. In the end, Hoddenbach had persuaded him not to avenge his nephew's murder. Torres had returned the guns and quickly left town. This was his first visit back to Chicago. "I felt like a punk," he told me, before transferring to the present tense. "I feel shameful." He said he had sought revenge for people who weren't related to him—"people who weren't even no blood to me." But he held back in the case of his nephew. "I still struggle with it," he said. On the ride over to the restaurant, Torres had been playing a CD of his nephew's favorite rap songs. It got him hyped up, and he blurted out to Hoddenbach, "I feel like doing something." Hoddenbach chided him and shut off the music. "Stop being an idiot," he told Torres.

"Something made me do what Zale asked me to do," Torres said later, looking more puzzled than comforted. "Which is respect my brother's wishes."

When Slutkin heard of Hoddenbach's intervention, he told me: "The interrupters have to deal with how to get someone to save face. In other words, how do you not do a shooting if someone has insulted you, if all of your friends are expecting you to do that? . . . In fact, what our interrupters do is put social pressure in the other direction."

He continued: "This is cognitive dissonance. Before Zale walked up to him, this guy was holding only one thought. So you want to put another thought in his head. It turns out talking about family is what really makes a difference." Slutkin didn't take this notion to the interrupters; he learned it from them.

ONE JANUARY NIGHT AT 11 p.m., Charles Mack received a phone call that a shooting victim was being rushed to Advocate Christ Medical Center. Mack drove the 10 miles from his home to the hospital, which houses one of four trauma centers in Chicago. Two interrupters, Mack and LeVon Stone, are assigned there. They respond to every shooting and stabbing victim taken to the hospital. Mack, who is 57 and has a slight lisp, is less imposing than his colleagues. He seems always to be coming from or going to church, often dressed in tie and cardigan. He sheepishly told me that his prison term, two years, was for bank fraud. "The other guys laugh at me," he said. LeVon Stone is 23 years younger and a fast talker. He's in a wheelchair, paralyzed from the waist down as a result of being shot when he was 18.

Advocate Christ has come to see the presence of interrupters in the trauma unit as essential and is, in fact, looking to expand their numbers. "It has just given me so much hope," Cathy Arsenault, one of the chaplains there, told me. "The families would come in, huddle in the corner and I could see them assigning people to take care of business." Mack and Stone try to cool off family members and friends, and if the victim survives, try to keep them from seeking vengeance.

The victim that night was a tall 16-year-old boy named Frederick. He was lying on a gurney just off the emergency room's main hallway. He was connected to two IVs, and blood was seeping through the gauze wrapped around his left hand. Mack stood to one side; Stone pulled up on the other.

"You know, the most important thing is—" Mack ventured.

"You're alive," Stone chimed in.

Stone then asked Frederick if he had heard of CeaseFire. The boy nodded and told them that he had even participated in a CeaseFire rally after a killing in his neighborhood.

"We try to stop violence on the front end," Stone told Frederick. "Unfortunately, this is the back end. We just want to make sure you don't go out and try to retaliate."

The boy had been shot—one bullet shattered his thigh bone and another ripped the tendons in two fingers. Nonetheless, he seemed lucid and chatty.

"My intention is to get in the house, call my school, get my books and finish my work," he told Mack and Stone. He mentioned the school he attends, which Mack instantly recognized as a place for kids on juvenile-court probation. Frederick told his story. He was at a party, and a rival clique arrived. Frederick and his friends sensed there would be trouble, so they left, and while standing outside, one of the rival group pulled a gun on them. Frederick's friend told him earlier he had a gun. It turned out to be braggadocio, and so when his friend took off running, so did Frederick, a step behind. As he dashed through a narrow passageway between buildings, he heard the shots.

"Can I ask why you're in the wheelchair?" Frederick asked Stone.

"I got shot 15 years ago," Stone told him. Stone didn't say anything more about it, and later when I asked for more detail, he was elusive. He said simply that he had gotten shot at a barbecue when he tried to intervene in a fistfight.

"You doing good," Stone assured him. "You got shot. You're here. And you're alive. What do you do when you get out of here?"

"You got to stop hanging with the wrong person, thinking you're a Wyatt Earp," Frederick said, speaking in the third person as if he were reciting a lesson.

At that point, Frederick's sister arrived. She explained that she was bringing up her brother. She was 18.

"He just wants to go to parties, parties, parties," she complained. "But it's too dangerous." She started to cry.

"Don't start that, please," Frederick pleaded.

Mack left a CeaseFire brochure on Frederick's chest and promised to visit him again in the coming weeks.

LAST MAY, AFTER A 16-YEAR-OLD boy was killed trying to protect a girl from a gunman on a city bus, Slutkin appeared on a local public-television news program. He suggested CeaseFire was responsible for sharp dips in homicide around the city. Slutkin, some say, gives CeaseFire too much credit. Carl Bell, the psychiatrist, was on the program with Slutkin that night. "I didn't say anything," he told me. "I support Slutkin. I'm like, Slutkin, what are you doing? You can't do that. Maybe politically it's a good thing, but scientifically it's so much more complex than that. Come on, Gary."

Last year, CeaseFire lost its $6 million in annual state financing—which meant a reduction from 45 interrupters to 17—as part of statewide budget cuts. One state senator, who had ordered an audit of CeaseFire (released after the cuts, it found some administrative inefficiencies), maintained there was no evidence that CeaseFire's work had made a difference. (The cuts caused considerable uproar:

The *Chicago Tribune* ran an editorial urging the restoration of financing, and the State House overwhelmingly voted to double CeaseFire's financing; the State Senate, though, has yet to address it.)

It can be hard to measure the success—or the failure—of public-health programs, especially violence-prevention efforts. And given Slutkin's propensity to cite scientific studies, it is surprising that he hasn't yet published anything about CeaseFire in a peer-reviewed journal. Nonetheless, in a report due out later this month, independent researchers hired by the Justice Department (from which CeaseFire gets some money) conclude that CeaseFire has had an impact. Shootings have declined around the city in recent years. But the study found that in six of the seven neighborhoods examined, CeaseFire's efforts reduced the number of shootings or attempted shootings by 16 percent to 27 percent more than it had declined in comparable neighborhoods. The report also noted—with approbation—that CeaseFire, unlike most programs, manages by outcomes, which means that it doesn't measure its success by gauging the amount of activity (like the number of interrupters on the street or the number of interruptions—1,200 over four years) but rather by whether shootings are going up or down. One wall in Slutkin's office is taken up by maps and charts his staff has generated on the location and changes in the frequency of shootings throughout the city; the data determine how they assign the interrupters. Wes Skogan, a professor of political science at Northwestern (disclosure: I teach there) and the author of the report, said, "I found the statistical results to be as strong as you could hope for."

BALTIMORE, NEWARK AND KANSAS CITY, Mo., have each replicated components of the CeaseFire model and have received training from the Chicago staff. In Baltimore, the program, which is run by the city, combines the work of interrupters and outreach workers and has been concentrated in one East Baltimore neighborhood. (The program recently expanded to a second community.) Early research out of the Johns Hopkins Bloomberg School of Public Health shows that in the East Baltimore neighborhood there were on average two shootings a month just before the program started. During the first four months that interrupters worked the streets, there had not been a single incident.

"My eyes rolled immediately when I heard what the model was," says Webster of Johns Hopkins, who is studying the Baltimore project. Webster knew the forces the interrupters were up against and considered it wishful thinking that they could effectively mediate disputes. "But when I looked closer at the data," Webster continues, "and got to know more about who these people were and what they were doing, I became far less skeptical and more hopeful. We're going to learn from it. And it will evolve." George Kelling, a Rutgers professor of criminal justice who is helping to establish an effort in Newark to reduce homicide, helped develop the "broken window" theory of fighting crime: addressing small issues quickly. He says a public-health model will be fully effective only if coupled with other efforts, including more creative policing and efforts to get gang members back to school or to work. But he sees promise in the CeaseFire model. "I had to overcome resistance," Kelling told me, referring to the introduction of a similar program in Newark. "But I think Slutkin's onto something."

Most of the police officials I spoke with, in both Chicago and Baltimore, were grateful for the interrupters. James B. Jackson, now the first deputy superintendent in Chicago, was once the commander of the 11th district, which has one of the highest rates of violent crime in the city. Jackson told me that after his officers investigated an incident, he would ask the police to pull back so the interrupters could mediate. He understood that if the interrupters were associated with the police, it would jeopardize their standing among gang members. "If you look at how segments of the population view the police department, it makes some of our efforts problematic," Baltimore's police commissioner, Frederick H. Bealefeld III, told me. "It takes someone who knows these guys to go in and say, 'Hey, lay off.' We can't do that."

Like many new programs that taste some success, CeaseFire has ambitions that threaten to outgrow its capacity. Slutkin has put much of his effort on taking the project to other cities (there's interest from Los Angeles, Oakland and Wilmington, Del., among others), and he has consulted with the State Department about assisting in Iraq and in Kenya. (CeaseFire training material has been made available to the provincial reconstruction teams in Iraq.) Meanwhile, their Chicago project is underfinanced, and the interrupters seem stressed from the amount of work they've taken on.

THE INTERRUPTERS HAVE CERTAIN understandings. At the Wednesday meetings, no one is ever to mention anyone involved in a dispute by name or, for that matter, mention the name of the gang. Instead they refer to "Group A" or "Group B." They are not investigators for the police. In fact, they go out of their way to avoid knowing too much about a crime. When Highsmith and Prater left me the night of the failed drug deal, they began working their contacts. Highsmith found someone who knew one of the stickup men and who, at Highsmith's request, negotiated with them. Highsmith's contact persuaded the robbers to return enough of the money to appease the drug-buyer's anger. When I met with the intermediary a few weeks after things were resolved, he was still stirred up about the robbery. "I was mad enough to do anything," he told me, making it clear that he and his friends had been hunting for the stickup guys. "This could've been a hell of a lot worse than it was." To this day, neither Highsmith nor Prater know the identities of anyone except the intermediary—and they want to leave it that way.

The interrupters often operate by instinct. CeaseFire once received a call from the mother of a 15-year-old boy who wanted out of a gang he joined a few weeks earlier. The mother told Hoddenbach and another interrupter, Max Cerda, that the gang members chased her son home every day from school threatening to beat him. They had shot at him twice. Hoddenbach found the clique leaders and tried to talk sense to them. If the boy didn't want to be in the gang, he told them, he'd be the first one to snitch. The gang members saw the logic behind that but insisted on giving him a beating before releasing him. Hoddenbach then tried another tack: he negotiated to let him leave the gang for $300—and no thrashing. The family, though, was only able to come up with $50, so Hoddenbach, Cerda and another interrupter came up with the rest. At their next Wednesday meeting, some interrupters were critical of Hoddenbach for paying what they considered extortion money. "It was kind of a messed-up way, but it was a messed-up way that works," Hoddenbach said.

It was nearly three months before Charles Mack could find time to visit Frederick, the young shooting victim. Frederick had since moved in with his great-grandmother in a different part of town. In his old neighborhood, he told Mack, "there always somebody who knows you. And I had a reputation." He complained to Mack that he had never been interviewed by the police but then declared he would never identify the person who shot him anyway. "I'm going to leave it alone," he said. As is so often the case, Frederick couldn't remember the genesis of the disagreement between his clique and the other. Mack promised to stay in touch, and as we dropped him off, Mack turned to me and said, "I think he's going to be all right." It sounded like both a proclamation as well as a hopeful aside.

Not long ago, I stopped by to visit with Hoddenbach at the Boys and Girls Club, where he holds down a second job. It was a Friday evening, and he was waiting for an old associate to come by to give him an introduction to a group of Hispanic kids on the far North Side. Apparently, earlier in the week, they bashed in the face of an African-American teenager with a brick. From what Hoddenbach could make out, it was the result of a long-simmering dispute—the equivalent of a dormant virus—and the victim's uncle was now worried that it would set off more fighting. As we sat and talked, Hoddenbach seemed unusually agitated. His left foot twitched as if it had an electric current running through it. "If these idiots continue," he told me, "somebody's going to step up and make a statement."

Hoddenbach also worried about Torres, who had recently gone back to Texas and found a job working construction. Hoddenbach says he originally hoped Torres would stay in Chicago and establish some roots, but then decided he'd be better off in another town. "I kept him out of one situation, but I can't keep him out of all of them," Hoddenbach said. This may well speak to CeaseFire's limitations. Leaving town is not an option for most. And for those who have walked away from a shooting, like Torres, if there are no jobs, or lousy schools, or decrepit housing, what's to keep them from drifting back into their former lives? It's like cholera: you may cure everyone, you may contain the epidemic, but if you don't clean up the water supply, people will soon get sick again.

Slutkin says that it makes sense to purify the water supply if—and only if—you acknowledge and treat the epidemic at hand. In other words, antipoverty measures will work only if you treat violence. It would seem intuitive that violence is a result of economic deprivation, but the relationship between the two is not static. People who have little expectation for the future live recklessly. On the other side of the coin, a community in which arguments are settled by gunshots is unlikely to experience economic growth and opportunity. In his book *The Bottom Billion*, Paul Collier argues that one of the characteristics of many developing countries that suffer from entrenched poverty is what he calls the conflict trap, the inability to escape a cycle of violence, usually in the guise of civil wars. Could the same be true in our inner cities, where the ubiquity of guns and gunplay pushes businesses and residents out and leaves behind those who can't leave, the most impoverished?

In this, Slutkin sees a direct parallel to the early history of seemingly incurable infectious diseases. "Chinatown, San Francisco in the 1880s," Slutkin says. "Three ghosts: malaria, smallpox and leprosy. No one wanted to go there. Everybody blamed the people.

Dirty. Bad habits. Something about their race. Not only is everybody afraid to go there, but the people there themselves are afraid at all times because people are dying a lot and nobody really knows what to do about it. And people come up with all kinds of other ideas that are not scientifically grounded—like putting people away, closing the place down, pushing the people out of town. Sound familiar?"

Jina Moore

Reading the Wounds

FROM *SEARCH*, NOVEMBER–DECEMBER 2008

Those fleeing torture and persecution in their home countries must submit to a medical examination in order to verify their claims as part of their bid to find asylum in the United States. But what are the lingering physical signs of torture? Jina Moore follows two doctors who have learned what torture looks like.

"Mr. A, I need you to take your shirt off." Dr. Rajeev Bais stretched a pair of latex gloves over his hands. He reached into his pocket and pulled out a paper tape measure, wound like a tailor's into a coil. Mr. A sat before him, on an exam table draped in crinkly white paper, at Elmhurst Hospital in Queens, New York.

"I'd like to measure the scars on your back and arms," Bais said, walking around the table. He moved like he talks, gracefully and deliberately, with an ease that inspires trust. Bais has brown eyes, and what's left of his black hair is buzzed short. Two gold charms his mother gave him rest near his collarbones: Ganesha, the Hindu elephant god, and Om, the symbol for the sound of the universe.

Mr. A silently slipped his arms out of his sleeves. Bais surveyed his upper body for evidence of cuts, deep bruises or other effects of physical abuse. He zoned in on the arms.

"There's some scarring here, near your elbows," Bais said, unrolling his tape measure. "How did that happen exactly?"

"That's where the cords were," Mr. A told him.

Bais wrote in his notebook: *5.5 cm by 1 cm linear shaped hyperpigmented scar just superior to the antecubital fossa.*

"What about your back?" he asked. "And your shoulder?"

Four small, dark ovals the size of nickels clustered around a thin, straight scar on the left side of Mr. A's lower back. Bais found two larger, misshapen circles, like stretched-out quarters, on Mr. A's left shoulder.

"The wooden sticks had sharp objects on them," Mr. A said.

3 cm by 3.5 cm, Bais wrote. *Raised.*

Bais catalogued six other scars of varying sizes and shapes on Mr. A's arms and back. When Mr. A removed his pants, Bais found more scarring on the thighs. He knelt to measure a long, uneven mark near Mr. A's left ankle.

"How did it heal?" he asked. "Was there any bleeding or pus?"

"It bled a lot," Mr. A answered. "And it swelled up."

"How long did it bleed?"

"A few months, but not all the time."

Bais had studied to be an emergency room physician—one of those George Clooney types, suave under pressure, whose days are built of short, intense encounters with car crash survivors, stabbing victims, and people with failing organs. But now he was encountering a kind of stress he had not learned about in medical school.

In their off hours, Bais and his colleagues become forensic physicians: interviewing asylum seekers who say they have been tortured, listening as they explain, over two or three or four hours, what happened to them back home. Then, with physical exams, Bais searches for the scars that can corroborate such stories. If, at the end of all this, the doctors believe the patients' stories, they write what may be the key to an asylum case: a medical affidavit, the closest thing to proof of torture there is.

Despite the many horror stories Bais has seen, he remains shocked that, actually, torture is rarely unique. There is a set of common practices, and those practices leave distinct interior and exterior scars. Bais and his colleague, Lars Beattie, have taught themselves what those scars might look like, and how to explain torture to judges, who grant the coveted right to remain in America only to refugees who can prove "a well-founded fear of persecution." The medical affidavits they write are based on a simple idea: that the body tells a story, one that might save a patient's life. The irony is that this is a modern inversion of precisely the philosophy that brought torture into the world in the first place: that truth can be coaxed from the body.

Bais made a few more notes and told Mr. A he could get dressed. Then he stepped out of the examination rooms and took a deep breath.

> *They just kept saying 'confess, confess. Confess to Kenya, confess to Riyadh.' I kept saying the Shahadah ["There is no god but Allah and Muhammad is his messenger"] and they kept beating me and mocking my religion. . . . They told me: "We're going to kill you and bury you here," and all the time I was wishing that they would.*
>
> —Abdul Rahman al-Yaf'i[1]

Torture has always been, in the words of one doctor, "a social institution," executed for the sake of a cause and on behalf of a people, by practitioners who see themselves as the guardians of those people. From its earliest documented inception, in ancient Greece, torture was also exclusive: it could be used against some individuals, but not all. "It is [for] the slave," Aristotle wrote, "and under certain circumstances, the foreigner."

It was never, in other words, for the vaunted Greek citizen, member of the legally protected population on whose behalf the torture was employed, usually in pursuit of criminal matters. This was no ticking-time-bomb scenario; in ancient Greece and Rome, torture was a method not simply for extracting information, but for purifying its source. Slaves and foreigners had to be made worthy of what we would today call offering testimony—by suffering for it.

Over time, the relationship between pain and proof took on a religious character. Before the development of a criminal justice system, the ordeal had become a common test of innocence. The assumption was that anyone telling the truth would be strengthened by God and survive; liars, abandoned by God, would get what they had coming.

The twinning of torture and truth found strength in another religious place: the confessional. In 1215, the Fourth Lateran Council made sacramental confession an annual Christian duty. At the same time, Europe was setting up its first modern legal systems, and confession seemed like the best kind of proof of a crime. Witnesses might have poor vision, faulty memories, or untoward motives, and juries could be equally untrustworthy. If self-incrimination was the prerequisite for spiritual absolution, why not also for guilty verdicts? Confession was embraced so fervently by the courts that it became known, in 13th century England, as "the queen of proofs"—and torture became her lackey, a tool officially sanctioned by the state.

If the world of the sacred inspired secular uses for pain, it dallied in endorsing them. Religious directives prohibited clergymen from using torture—but not from cooperating with people who were not so fettered in their techniques. To combat religious dissent, the church relied on the state to call it a crime. Europe's major monarchs outlawed heresy, and the church handed over—rendered, one could say—cases of heresy to jails and courts whose rules of engagement included torture. Eventually, the church eliminated the ban on torture, laying the foundation for the Inquisition.

In France, by the 17th century, torture had become a way of physically uncovering the truth. French jurists believed one's guilt or innocence could be made manifest in the tortured body of the accused. Rather than coercing confessions from criminals, torture literally spoke for them by leaving injuries whose signs experts could dissect.

It is into this long and tangled relationship between pain and truth that Rajeev Bais and Lars Beattie stepped three years ago, when they opened the volunteer-run Libertas Human Rights Clinic in Queens. Torture survivors come here for free basic medical care and the affidavits they need for their asylum claims. So far, the doctors have treated and written affidavits for nearly one hundred people (all of whom, as far as they know, have received asylum).

Medical studies say up to 35 percent of refugee patients treated by doctors are likely to have survived torture. An even higher proportion may come through Elmhurst, a public hospital in the country's most diverse county; 46 percent of Queens residents are foreign-born. "If there's a place to look for these people," Beattie says, "it's Queens."

The Libertas Clinic and its doctors are part of a loosely-linked world of medical professionals who write affidavits in asylum cases. The medical community first began to notice a proportionally high number of torture survivors among their patients in the late 1970s and 1980s, when refugees fled violence across Central America. In the 1970s, Amnesty International launched an anti-torture campaign, which opened up a conversation in the medical community about how to help survivors.

In 1982, the world's first center for treating torture survivors was established in Denmark. Three years later, the first U.S. center opened in Minneapolis, which has been a hub for refugee resettlement for the last thirty years. By the 1990s, practitioners on both continents were developing specialized treatment techniques for torture survivors, and researchers began investigating the physiology and biochemistry of the body's experience during torture.

Today, the National Consortium of Torture Treatment programs links almost forty health care centers, advocacy organizations, human rights clinics and other organizations, like Libertas, working on behalf of torture survivors in the United States. The International Council for the Rehabilitation of Torture Survivors re-launched the professional journal *Torture* three years ago, at the same time it released a study on medical literature published about torture since 1998. "Perhaps the most important finding is that either torture has increased worldwide," the study said, "or exposure of torture events has improved." Whatever the cause, requests for affidavits to members of Physicians for Human Rights' Asylum Network, a collection of four hundred volunteers like Bais and Beattie operating independently in thirty cities, have been increasing for the last four years.

The problem, these doctors have long known, is that there is no instrument for conclusively proving someone's case. "In order to establish the truth or falsehood of the allegations," notes from an Amnesty International medical conference in 1978 say, "it would help immeasurably to have available standardized tests that can be easily applied. Physical proof would establish beyond doubt that the alleged torture has in fact occurred."

Medieval beliefs about the body's ability to showcase the truth are not, it turns out, so far from modern medicine. In these cases, the allegations are of a patient's innocence, not a criminal's guilt, and the relationship is more direct: the question is not what the body says a person has done, but what the body says about what has happened to the body. It's a welcome tautology.

Each patient who visits Libertas requires a half-dozen or more hours of patient listening and observation. By the time Mr. A jumped up on Bais' exam table, he had already told Bais he was tortured three times, beaten with wooden sticks, kicked by men wearing military boots, hung upside down from his elbows. But he'd also told Bais about his life before torture, about his new life in America, about the last movie he'd seen. That kind of small talk helps patients feel like their lives are normal. And their doctor is normal, too.

"We'll talk about anything," Bais said, "to show them I'm not some freak really into looking at torture scars."

> Falanga *is an overwhelmingly powerful force that works on your whole system. You have the impression of sliding down a vast, shining slope, then you're flung into a hard granite wall. If you didn't know they were beating you on your feet it would be impossible to make out where it was coming from. You can see the movements of the torturer. The strokes of the rod are the granite wall. The slope is the interval between each stroke. When there is a regular rhythm to it it's less painful than when it's irregular. They are well aware of this subtle variation and they hit you fast and slow, alternately. They beat from heel to toe and back again. They know your first reaction is to arch your feet. They don't mind you doing this as they know quite well that after a dozen blows your foot will have swollen so much that it fills up your shoe.*
>
> —Pericles Korovessis

Bais had to become as much an expert in torture techniques as those who practice them. It was a lonely, self-driven study, during which he relied on the field's canonical texts: the Istanbul Protocol, the 1998 United Nations manual on how to conduct interviews and physical and psychological exams of survivors, and *Examining Asylum Seekers*, a manual for physicians and psychologists written in a dissonant combination of medical jargon and torture slang. The "parrot perch," it explains of the nickname for a kind of suspension torture, is "the suspension of a prisoner by the flexed knees with a bar passed below the popliteal region, usually with the wrists tied to the ankles. It may produce tears in the cruciate ligaments of the knees." Which is a specialist's way of saying if you find tears in the cross-shaped tissue that keeps the shinbone and the calf bone from banging into each other, you've probably got a patient who was bound and hung from his knees over a rod.

These books give best-case diagnostic scenarios, which require no specialization for diagnosis. Electrocution, for example, is straightforward. What happens to the body when voltage rushes through it is pretty universal: muscles cramp and form scarred bundles, and raised scars often appear where the current entered the body. But diagnostic proof, medical science's version of "beyond a reasonable doubt," doesn't last forever. After electrocution, calcium deposits on the outermost layers of skin—deposits that get there only when current passes through—may linger, but not longer than a week. Even on patients who have endured severe trauma, treatment books warn, doctors are unlikely to find diagnostic evidence.

The best doctors like Bais and Beattie can do is look for the kinds of marks they know specific torture techniques might leave. Pounding the ears with cupped hands, a method known as *telefono*, might pop an eardrum, leaving scar tissue. Beatings on the soles of the feet—*falanga*—might kill the heel's soft tissue. But it's also possible not to find anything. Unless a bone breaks, a joint pops, or nerves and muscles are damaged, the body keeps silent about its trauma.

Torturers know this. They also know putting a shoe on someone's foot before beating it with a lead pipe spreads out every blow, lessening the chance of destroying tissue, and that putting a wet towel between the skin and, say, a cattle prod, usually prevents the electrical current from leaving a long-lasting mark. Suspension hanging—dangling someone from their ankles or wrists or knees, known during the Inquisition as the "queen of torments"—is infamous for its absence of physical evidence. That's precisely why it is practiced around the world.

There may be, some doctors think, an underground conversation happening about how to cause pain without scarring.

"Somebody somewhere said, 'This is a good way to make people suffer,'" Beattie said. If so, the conversation seems to be regional. *Telefono* is popular in Peru but almost unheard of in Iran, where *falanga* is something of a standard operating procedure. Near-suffocation with water is so common in Latin America that it is known, among specialists in the U.S., by an English and Spanish hybrid: "wet *submarino*."

Given the level of sophistication torturers practice, Beattie wasn't surprised that most of the physical evidence of torture had disappeared from the body of Mr. Y, one of his earliest patients. Mr. Y was arrested in 1995 and imprisoned in Chechnya, where police officers tightened plastic bags around his head until he nearly suffocated, according to his interview with Beattie. They beat him with soda bottles, one-third full of water, a half-dozen times a day. At one point, they tied his wrists behind his back, pulled his arms over his head, and hung him from the ceiling. And they used something like a water-dropper to apply a liquid that made his back feel "like it was burning."

But few scars are left. Beattie found clusters of warts on his neck and chest. He measured them and felt them with his fingertips. Then he wrote: *Multiple small 1-2 cm diameter raised hard keloids on the neck and torso. They are raised and rubbery in texture. There is no surrounding erythema or induration. The lesions are not fluctant.*

In other words, the thick, circular scars are still the same color as his skin, which means the capillaries weren't permanently damaged by whatever caused the scarring. They haven't hardened permanently, and they don't shift under pressure. Based on that evidence, Beattie concludes, Mr. Y was likely burned, probably with an acid or a base—just as he had told Beattie.

At this point, when the scars confirm the story, Beattie usually writes the first draft of his affidavit. But with Mr. Y, he hesitated. He concedes that what happened to Mr. Y sounded pretty bad, and yet: "I mean, the guy is from the Caucasus," Beattie said later. "But the Chechens are no angels either. There's torture on both sides." He paused. "So is this guy really a good guy to let into the country?"

> *It was as if my arms were coming off. . . . The pain became so bad that my screams drowned [the torturers'] voices. . . . Even when they stopped torturing you physically, the screams of others began to torment you psychologically. After a while, I was able to pick out which torture was being applied from the screams.*
>
> —Sema Ogur

Lars Beattie would never describe himself as powerful. He's six feet tall and still, at 40, a little gangly, with a scruffy goatee, wide, green eyes and sandy hair that flops over his eyebrows. He speaks softly, in six- or seven-word spurts, and turns his eyes down when he has to say gruesome things like, "They burned him with a hot machete on his thigh." When he's not stitching people up in the ER, or at the clinic, Beattie paints, or makes furniture, or tries to be a good dad.

Beattie resists being characterized as powerful; the judge, after all, decides who stays in the United States and who's deported, and Beattie has never even met an asylum judge. When pressed, he conceded only, "I guess I make some sort of decisions about people."

It is, often, on the basis of the decisions he makes that a refugee like Mr. Y is allowed to stay in America. An affidavit carries a great deal of weight in the asylum process. To win asylum, a person has to prove that going home threatens his life, by virtue of his membership in an at-risk social group. Some risks are clear—think of sending Kurdish refugees back to Saddam Hussein's Iraq—but most require stacks of substantiating paperwork.

The medical affidavit is often the most powerful in that stack because it functions as tacit proof of torture. Essentially, the doctor who writes it is saying, "I believe this guy," though no Libertas Clinic affidavit uses that vocabulary. The most the doctors there will do is lay out two parallel stories, one told by the torture survivor and one by his or her body. When the scars and stories match, they write, "The scarring is consistent with the mechanism of injury."

The key to the match is how a scar heals. It's hard to mistake the raised, misshapen circular scar of a cigarette burn, for example, for anything else; nicotine ruins its edges and leaves the skin looking curdled. Gashes left by whipping give clues about how much force went into the abuse; internal scar tissue suggests, by its position and intensity, its perpetrator.

Beattie acquired most of his knowledge observing scars from ordinary medical resources: on-the-job experience dealing with accidental injuries he sees in the ER, and med school textbooks. He swears by *Wounds and Lacerations*, which describes how virtually every kind of abrasion looks when it heals well and when it heals poorly. Knowing the difference helps him figure out who's telling the truth. "It's hard to fake how a knife cuts you," he said, "the way it comes in, how it heals."

Bodies, in other words, can expose a lie. One doctor in Manhattan interviewed a man who claimed the scars above his nipples were the result of torture he suffered in Liberia under Charles Taylor's regime. But the scars were symmetrical and equidistant, implying a precision which raised the doctor's suspicion. A Liberian culture organization to which he described the markings told him the scars were more likely evidence of a tribal ritual ceremony.

After dozens of affidavits—a nineteen-year-old woman from Mali with female genital mutilation; a Tibetan tortured with electric cattle prods fifteen years ago; a man from the Ivory Coast burned with a machete and hung from his wrists—Beattie has learned the signs of a well-planned lie. Usually, how the patient tells a story tips off what Beattie calls "the gut gestalt," which he's honed after seeing so many schemers in the ER.

"I think people imagine torture in a comic book type way. They say, 'Oh, look at this scar here. And I've also got this one!'" he said. "People who've gone through this generally have a huge psychological burden to get off their chest. They say, 'Well, it all started when . . .' They try to work through what happened to them in a storybook kind of way. And you're emotionally invested."

But sometimes, as in Mr. Y's case, the emotional investment doesn't seem to be quite enough. Beattie never doubted his story, but he did wonder about the moral equivalence of pain: Who's to say, he wondered, that Mr. Y never inflicted brutality on someone else? There's no room in legal proceedings for that kind of question, though; the standards of proof in a courtroom, rather than the questions of Beattie's conscience, dictated his decision.

"I decided, look, I believe him as a human being. I believe he was victimized there," Beattie says. "And I believe I'm doing the right thing by writing the affidavit. But you know, I'm sure—" He paused. "Other people might feel otherwise."

> *One practice . . . was "short shackling" where we were forced to squat without a chair with our hands chained between our legs and chained to the floor. If we fell over, the chains would cut into our hands. We would be left in this position for hours before an interrogation, during the interrogations (which could last as long as twelve hours), and sometimes for hours while the interrogators left the room. The air-conditioning was turned up so high that within minutes we would be freezing. There was strobe lighting and loud music played that was itself a form of torture.*
>
> —Shafiq Rasul and Asif Iqbal

In the United States, there has been of late little agreement about what torture actually is. Part of the disagreement stems from the epistemological problem of pain. Severe pain confounds language; it leaves its sufferer grasping for similes, metaphors and analogies that can only approximate the feeling and intensity of pain. "To have great pain is to have certainty," writes Elaine Scarry, a literature professor at Harvard University, in her landmark book *The Body in Pain*. "To hear that another person has pain is to have doubt."

Much of that doubt, these days, is political. The debate in America about what counts as torture is rarely about what actually happens to the body and more about whether the end, the extraction of vital national security information, is justified. We avoid thinking about what "short shackling" means and concentrate instead on what we hope it will get us: What if the shackled suspect knows something about the planning of the September 11 attacks? What if he knows but refuses to name someone who wants to set off a dirty bomb? What if, in fact, that dirty bomb is out there somewhere, its ignition clock ticking away the seconds until explosion, while we dither about the Geneva Conventions?

In his memo offering the Bush Administration the leeway it sought to undertake harsh interrogations, extraordinary renditions, and other now familiar practices on a contested continuum of torture, former Justice Department Office of Legal Counsel Attorney John Yoo wrote, "Torture is not the mere infliction of pain or suffering on another. . . . The victim must experience intense pain or suffering of the kind that is equivalent to the pain that would be associated with serious physical injury so severe that death, organ failure, or permanent damage resulting in a loss of significant body function will likely result." It doesn't count as torture, the U.S. government seemed to say, unless it almost kills you.

Often the only people standing between the "mere infliction of pain and suffering" and near-death torture are, ironically, doctors. Steven Miles, a professor of internal medicine at the University of Minnesota, combed through 35,000 pages of government documents released under the Freedom of Information Act to understand the role medical personnel had played in military interrogations in Guantánamo, Iraq and Afghanistan.

"They were supposed to make sure this man has his medications available during harsh interrogation," he said, "make sure his blood pressure stays under such and such—that kind of thing." In his 2006 book, *Oath Betrayed: Torture, Medical Complicity and the War on Terror*, he quotes from directives he found by former Secretary of Defense Donald Rumsfeld insisting harsh interrogations take place only with "the presence or availability of medical personnel." Miles also found more than two hundred military studies, by his count, that concluded the intelligence elicited by torture is usually faulty.

Compartmentalizing torture into medically measurable units—pulse rate, blood pressure level, lung capacity, brain activity—suggests that even the most extreme of physical experiences can be approached in a routine way. And in fact, the physical experience of torture is, in some ways, generic. Skin doesn't burn any faster or more deeply in a prison cell than on a hot stove, and its wounds don't heal any more quickly if they weren't deliberately inflicted.

What makes torture a unique kind of pain—what makes torture torture—is the purpose of the person committing the abuse. It is, in the words of one treatment center, "the intentional and systematic infliction of physical or psychological pain and suffering in order to punish, intimidate, or gather information." In other words, the torturer has to want something. This isn't so different than the circumstances in which torture first sprung up, hundreds, perhaps thousands, of years ago. Then, as now, torture was the infliction of suffering on an individual in service of a community. And as soon as pain becomes an exercise in political power, the debate about the definition of torture is over.

> *I am just like a chook [small aquatic plant] in the middle of the river or the sea. My life is empty. Every move I make, sitting, sleeping, walking and standing, I feel as if I am all alone. . . . This feeling is still with me. But I try to forget about this feeling. . . .*
>
> —"CL"

Some patients who come to the Libertas Clinic have reassembled their lives; others are still fragile, years after their experience. There's no manual that tells Rajeev Bais or Lars Beattie how quickly their patients "should" heal; like the physical scars they seek, the mental and emotional scars of torture can linger, seemingly silent.

In fact, how quickly survivors recover might depend on why they think they suffered. Studies have found that torture survivors who believed they were beaten in the name of a noble cause more easily readjust to ordinary life than those who were arrested in a random sweep or a case of mistaken identity. "Two people can feel the same amount of pain and suffer to different degrees," says Dr. Frank Vertosick, a neurosurgeon and author of *Why We Hurt: A Natural History of Pain*. "It's the Joan of Arc phenomenon. I know I'm doing this for a cause, so the pain is just the pain."

The cause gives the experience a greater meaning that not only helps people recover from torture, some doctors say, but also to live through it in the first place. It's less about what's felt and more about how it's perceived.

Vertosick says pain is a sensation, like noise or taste or sound. Its signal is interpreted by the thalamus at the base of the brain, and, on its own, it means nothing to the body. But when the signal travels to and is interpreted by the brain's frontal lobes, the physical experience the body is having becomes what we describe as painful. Ultimately, the brain decides that pain hurts, and how much it hurts. Whether it "hurts more" to be hung from the wrists or to be forced to squat for hours at a time—both considered categories of suspension torture—depends on how the person being suspended interprets the experience.

"Some pain is positive—ritual pain, for instance. It has a positive meaning. So why is torture so harmful?" asked Victor Iacopino, a senior medical advisor to Physicians for Human Rights who wrote the Istanbul Protocol. "People are harmed because of the meaning torture has to them. Because they had to betray loved ones, friends, colleagues in order to survive. Because it's degrading."

Almost all of the patients who have come to the Libertas Clinic suffer from post-traumatic stress disorder. Almost all suffer from depression. Most tell their doctor that they're uninterested in the world or that they can't trust it. Lars Beattie refers them to psychologists, but he wonders how they ever really recover.

"A burn heals, the pain goes away. But if somebody took my hand and held it to the stove," he said, "if somebody did that to me, it's going to hurt a lot longer."

The seeming impossibility of recovery is what makes his patients' resilience so incredible to Beattie. He has heard enough gruesome details to trouble his subconscious for the rest of his life, but there's one in particular he remembers, one which resurfaces, over and over: a West African man was locked, naked, in a metal shipping container for fifteen days. His wrists were bound to his ankles, his thighs burned with a hot machete. Until he managed the unthinkable.

"He escaped. He walked without clothes to the border . . . and he manages to get here, with nothing," Beattie said with wonder in his voice. "He's gone through all this . . ." The doctor stopped and shook his head at the utter implausibility of what he was about to say. "And every day he gets up and sells his wares. He gets on with his life."

[1]*The voices of torture survivors woven through this article have been gathered from around the world over the last forty years.*

Abdul Rahman al-Yaf'i survived four months in prison in Jordan after his extraordinary rendition by Egypt, possibly in collaboration with the United States; his story appears in Amnesty International's April 2006 report, *USA: Below the Radar—Secret Flights to Torture and Disappearance.*

Pericles Korovessis was tortured by the Greek junta in 1967; he tells his story in his 1969 book, *The Method.*

Sema Ogur was a student detained twice by the Turkish government for a total of forty-seven days. Her husband was also arrested and tortured; her story appears in Bulletin no. 1, 1984, of Amnesty International's Campaign to Abolish Torture.

Shafiq Rasul and Asif Iqbal, both British nationals, issued a joint statement describing their imprisonment by American authorities at Guantánamo Bay, Cuba, in an open letter to the U.S. Armed Services Committee in 2004.

JENNIFER KAHN

A Cloud of Smoke

FROM *THE NEW YORKER*

When a first responder to the 9/11 attacks dies nearly five years later of lung disease, he is hailed as a hero who succumbed to the toxic fumes he inhaled during the rescue effort. But then the medical examiner of the City of New York announces the results of his autopsy. Jennifer Kahn covers the furor that erupts.

IN JANUARY, 2006, SHORTLY AFTER JAMES ZADROGA DIED at his parents' house, in Little Egg Harbor, New Jersey, Dr. Gerard Breton received a call from the Ocean County Medical Examiner's Office. A retired hospital pathologist, Breton now conducts autopsies for the county on contract. Compared with its more violent neighbors—Newark, to the north, and Atlantic City, to the south—the central New Jersey shore tends to generate uncomplicated deaths:

heart disease, liver failure, sunstroke. Zadroga's death, at the age of thirty-four, was different. Medical records showed that his health had been failing for almost five years, apparently owing to degenerative lung problems, which had forced him to go on sick leave from his job as a New York City homicide detective in 2002. He had spent the last year of his life dependent on a portable oxygen tank. After Zadroga died, on January 5, 2006, his father, Joe, requested an autopsy.

At the morgue, Breton surveyed the body. He noted two small bruises on the chin and a pair of faint scars on the abdomen—from stab wounds, he speculated—along with several tattoos. Cutting inward from the armpits to the pubic bone, he folded back the muscle and removed the rib cage with a saw. Six feet tall and weighing two hundred and sixteen pounds, Zadroga was solidly built, despite losing weight rapidly in the final months of his life. Even so, Breton recalls being struck by the size of Zadroga's lungs, which had a reddish, meaty appearance, and were unusually firm to the touch. "Normally, the lungs feel soft and spongy," he said. Lifting the lungs onto a scale, he also found that they were exceptionally heavy—almost three times the usual weight.

A dissection revealed the cause. Zadroga's lungs contained a vast number of foreign-body granulomas: knots of scar tissue that build up around inorganic particles, like the pearling around sand grains in an oyster. The scarring was so extensive and severe that the right ventricle of Zadroga's heart had thickened from the strain of trying to force blood through the ravaged vessels and capillaries. "It's not unusual to find a few granulomas in a lung," Breton noted later. "But usually this is something you see here and there." The granulomas in Zadroga's lungs were all over.

Breton couldn't identify the gritty particles lodged in the tissue with his microscope, though he remembers thinking that some of them looked "dustlike." He sent a sample to the Armed Forces Institute of Pathology, which identified the foreign material as a combination of talc, cellulose, calcium phosphate, and methacrylate plastic. Breton also spoke to Joe Zadroga, who said that his son's health had begun to deteriorate in the weeks following the World Trade Center attacks. Shortly after joining the recovery effort at Ground Zero, Joe said, James had developed a persistent cough.

After talking with Joe, Breton had little doubt about the underlying source of Zadroga's lung trouble. In his final report, he concluded that James Zadroga had died from respiratory failure due to severe panlobar granulomatous pneumonitis. He added that he felt "with a reasonable degree of medical certainty that the cause of death in this case was directly related to the 9/11 incident."

NEWS OF THE DIAGNOSIS SPREAD quickly. The hazy air in which rescue teams worked was already suspected of contributing to everything from chronic coughs to cancer, and Zadroga's death was widely seen as the harbinger of an epidemic. A few weeks later, the senators for New York and New Jersey—Charles Schumer, Hillary Clinton, Robert Menendez, and Frank Lautenberg—wrote to George Pataki, who was then the governor of New York, urging him to extend the list of 9/11 victims, and to compensate the families of first responders who became chronically ill after working at Ground Zero. The letter called attention to the predicament of Zadroga's four-year-old daughter, Tyler Ann, now orphaned—Zadroga's wife, Ronda, had died the previous year—and in the care of her father's parents. "Detective Zadroga served with great courage in the line of duty and paid the ultimate sacrifice for his heroism," the senators wrote, adding that his death "will not be the last to be suffered by the brave Americans who rushed to Ground Zero in the hours and days after September 11."

The so-called Zadroga Act, which Pataki signed in August, 2006, provided generous benefits to the families of city workers who died of 9/11-related illnesses. Tyler Ann became one of the first recipients. But there were other complications to come. In order for Zadroga's name to be added to the official victims' list—and, ultimately, engraved on the National September 11 Memorial—his status as a casualty of 9/11 had to be verified. In the summer of 2007, copies of

Zadroga's autopsy report and medical history (and, later, slides of tissue taken from his lungs) were sent to the Chief Medical Examiner of New York City, Dr. Charles Hirsch. In October, he returned his verdict. Calling his assessment "markedly different" from Breton's, Hirsch disputed the claim that the talc and the cellulose found in Zadroga's lungs came from Ground Zero. Instead, he said, the embedded material was pharmaceutical debris produced by injecting a solution of crushed prescription pills.

The news provoked an uproar. The *Daily News* ran an editorial accusing Hirsch of "smearing" Zadroga's reputation. The president of New York's Detectives' Endowment Association suggested that the city was trying to arm itself against pending class-action suits from Ground Zero workers. A congressional delegation demanded that an independent panel be established to investigate all deaths connected to dust exposure at the 9/11 sites. "The history of 9/11 should not be decided behind closed doors by one person," Carolyn Maloney, a New York congresswoman, announced. Michael Moore took up the cause on his Web site.

At a conference a few days later, Joe Zadroga appeared with Michael Baden, a celebrity forensic pathologist who had used one of Zadroga's lung-tissue slides for a segment on 9/11-related illness in his HBO series, *Autopsy: Postmortem with Dr. Baden.* Baden, a former Chief Medical Examiner of New York City who is now the forensic pathologist for the New York State Police, announced that he had reviewed the evidence and had no doubt that James Zadroga had died as a result of exposure to toxic dust at Ground Zero. There was no sign of any drug addiction, Baden said. "You could see glass fibres in there. You don't get that from injecting drugs."

In the months since the announcement, Baden has not changed his view. Discussing the case with me in his apartment, eighteen floors above the Museum of Modern Art, he acknowledged that granulomas can result from injecting ground-up medication—as happens when heroin users inject a solution of crushed methadone tablets—but disputed the idea that the wreckage in Zadroga's lungs could have been caused by shooting up. "People who have chronic lung disease from drug abuse have long histories of drug abuse, rehabilitation, doctors going in and out," he said. "You can't hide that sort of drug abuse." He had noted earlier that Breton had observed no needle scars on Zadroga's arms. "If I saw needle-track scars, that would make me think right away: intravenous abuse."

Baden would not speculate on what lay behind Hirsch's finding, though he wondered whether the case had involved something political. Either way, he said, he found the assessment baffling. "I think that Dr. Hirsch seems to have been misled by the fact of finding talc and cellulose, which can be found in people who grind down drugs that are meant for oral consumption. But they can occur in other ways." He said that he considered the assessment not only medically wrong but also "nasty"—a mean-spirited swipe at a guy "who was at Ground Zero, who really worked hard on the pile."

THE OFFICE OF THE CHIEF Medical Examiner has the unrenovated pallor of a forgotten city agency. Dimmed by a concrete overbite, the street entrance manages to look at once ominous and shabby—a homely approach to an agency that houses one of the largest and busiest forensic labs in the country. Even by the standards of other big cities, New York has a prolific capacity to produce dead bodies, and, as Chief Medical Examiner, Charles Hirsch is responsible for the processing of some twenty-five thousand fatalities a year—nearly half the city's annual total. Roughly fifty-five hundred of those cases require autopsy, including all deaths that are violent, sudden, mysterious, or in some way related to public or consumer safety—an infant asphyxiated by a car seat, say, or a runner who died from applying too much topical muscle cream.

Hirsch's role is covertly powerful. Medical examiners are frequently required to testify in criminal cases, and in the past few years he has overseen the investigation of several high-profile fatalities, including the sudden death of Heath Ledger and the fatal child abuse of Nixzmary Brown. Being the final medical authority in such cases—particularly those involving police brutality or city negligence—can be politically fraught, and, in New York, administrative infighting and sheer population density combine to create a singular institutional challenge.

For someone occupying such a controversial position, Hirsch has managed to remain unusually invisible—in part because, unlike many medical examiners, he rarely speaks with the press. In the course of his nineteen-year tenure, he has never been interviewed on television, and has hardly ever been quoted in print. Jonathan Hayes, a senior medical examiner who has worked in the office since 1990, told me that one of Hirsch's maxims is "Medical examiners get into trouble by saying too much, too soon, or to the wrong person."

Privacy, however, can also create problems. By refusing to detail the evidence supporting his interpretation of Zadroga's death, Hirsch appeared secretive, fostering suspicions about his motives and the validity of his verdict. Policemen in particular felt betrayed. Barbara Butcher, who coördinates crime-scene investigations between Hirsch's agency and the Police Department, said, "There are cops I've known for years who wouldn't talk to me about it."

When I visited Hirsch recently, he had, as usual, been in his office since seven in the morning. (His departures are less consistent. He lives with his wife just a few blocks north of the office, and Hayes told me, "I've looked up from bodies and seen Dr. Hirsch standing there at three in the morning.") Hirsch is seventy-one years old. Courteous and mannerly, he walks with a slight forward cant—the lingering effect of an operation to repair a herniated disk. Although his face is lean, the cheeks are rosy, with large ears and dark eyebrows framing a nose that is slightly off center. Before coming to New York, Hirsch went to school in Chicago and worked in Cleveland, and his habits still reflect Midwestern tastes: sober suits worn with narrow suspenders, and, for pleasure, a pipe of Captain Black tobacco. When making a point, he occasionally taps one finger lightly on the desktop. His speech is so soft that one has to strain to hear it.

At an afternoon meeting, conducted with ten of the agency's thirty forensic pathologists, Hirsch received brisk updates on a dozen ongoing investigations, including the apparent suicide of an eighty-year-old man, and the autopsy of a woman who was found mummified in her apartment wearing nothing but a sweater. Hirsch listened to these recitations quietly, removing and folding his glasses with care. Dissatisfied with the evidence presented in the suicide, he asked about the meaning of a note found on the man's kitchen table, which was addressed to relatives in Eastern Europe and ended, simply, "Farewell."

"Was the letter dated?" he asked. "Do you know how long it was sitting there?"

He also lingered on the mostly full bottles of prescription medication that were found in the room. "People who want to kill themselves tend to empty every bottle," he noted. A discussion ensued. A medical examiner pointed out that the only empty bottle recovered from the scene contained Ambien, a sleep aid—but that the toxicology panel had found no traces of Ambien in the body. The bathroom cabinet contained an unfamiliar antibiotic, possibly bought in Eastern Europe, and not listed in any of the standard pharmaceutical resources. Hirsch listened politely until the conversation ended. Then he turned to address the medical examiner in charge of the case. "The note isn't compelling, and the toxicology is borderline at best," he said quietly. "If you're going to conclude that this is an intentional self-destruction by medication, you're going to have to prove it."

Hirsch is known for his willingness to challenge seemingly obvious conclusions, a tendency that, in politically charged cases, has often made him unpopular. In 1989, a twenty-five-year-old black man named Richard Luke died in police custody after being physically restrained by the arresting officers. Although the medical examiner's office confirmed extensive bruising, Hirsch ruled that Luke died not as a result of police brutality but from acute cocaine intoxication—a verdict that prompted street protests. Hirsch insists that he has been unaffected by the excoriation following his findings on the Zadroga case but noted that he had received requests from several senators urging him to change his assessment. "It's the only instance in which political pressure has been exerted on me," he remarked tartly.

SINCE THE CONTROVERSY, HIRSCH HAS revealed little about the evidence in the case, but when we spoke he stressed the fact that the granulomas in Zadroga's lungs were concentrated in the arteries and not in the alveoli. Drug users who inject crushed pills routinely develop granulomas, because insoluble fillers used to bind a pill's active ingredient don't get digested—as they would if a pill was swallowed—and, instead, circulate through the bloodstream, eventually lodging in the narrow pulmonary arteries. The condition is well known and is sometimes referred to as "mainliner's lung." When I spoke to Michael Baden, he argued that Zadroga's lungs were in such bad shape that it was impossible to tell where the damage had originated. Hirsch disagreed. Crystals lodged in the pulmonary arteries can sometimes get pushed into the alveoli, he observed, because of the constant pressure of surging blood, but the same force does not exist in reverse. Therefore, the matter in Zadroga's pulmonary arteries must have entered his lungs through the bloodstream, not through inhalation.

The type of material found in Zadroga's lungs was also significant. Talc and cellulose are widely used pharmaceutical ingredients, Hirsch noted, but neither was considered a major respiratory hazard at Ground Zero: talc because it wasn't found in the air in abundance; cellulose because it is not considered particularly hazardous. (The molecules are too large to pass through the elaborate filtering apparatus of the upper airway.) The absence of needle marks Hirsch deemed insignificant, because clean needles cause less scarring.

Talking about the case later, Jonathan Hayes remarked that one of the dangers of forensic pathology, particularly in dramatic cases, is that people see what they expect to see. "I think there's a possibility that that's what happened here," he said. "This guy's been down in the World Trade Center, spent hours after hours on the pile. You look at his lungs: oh, they're full of scar tissue and inflammation!" He went on, "As Dr. Hirsch would say, 'Sometimes it's necessary to slay a beautiful theory with an ugly fact.' "

Interestingly, Hirsch was himself exposed to the miasma of debris created during the 9/11 attacks. Early that morning, he rounded up a small crew of examiners and drove to the base of the burning towers to scout sites for a temporary morgue. With an investigator named Diane Crisci, he walked south along the West Side Highway and had just reached the pedestrian walkway connecting the World Trade Center to the Winter Garden when the south tower collapsed.

The blast threw Hirsch underneath the walkway, where he was battered by flying rubble but sheltered from the worst of the falling debris. When he opened his eyes, Hirsch recalled, the air was so black that he couldn't tell whether he was on the surface or buried. He spat out the dust and located Crisci. Together, the pair hobbled away from the devastation—Hirsch with a sprained ankle and a gashed wrist, Crisci disabled by a broken leg. Hirsch arranged medical care for Crisci, then hitched a ride back to the office, his arm still bloody and his clothes covered in dust.

When I asked Hirsch whether he had experienced any symptoms from this exposure, he was emphatic that he had not. When I persisted—did he, at least, worry about the long-term effects?—he grew terse. "There's no point in worrying about things that you can't control," he said.

For professional reasons, Hirsch said, he continues to follow the epidemiological research, particularly a recent study that showed a statistical increase in sarcoidosis—an inflammatory disorder that attacks the lungs and other organs—among firefighters who worked at Ground Zero when the smoke was thickest. Last May, the weight of this study persuaded him to revise his opinion of five years earlier on the death of Felicia Dunn-Jones, a forty-two-year-old lawyer who died five months after being trapped in the initial explosive plume. In his review, Hirsch noted that, while the sarcoidosis that killed

Dunn-Jones may have been present before 9/11, the illness had almost certainly been exacerbated by the particulate-laden air she inhaled on the morning the towers fell. The ruling made Dunn-Jones the first dust casualty to be entered on the official list of 9/11 victims. It also prompted a number of other families to resubmit cases that had been closed.

Each of these claims will be examined, Hirsch told me, and as more information becomes available some deaths may eventually be reclassified. But the results may be slow in coming. Hirsch said that he'd had letters from survivors who had developed cancers of one form or another, and that he could only reply that the scientific evidence isn't there yet. "It's going to take a lot of epidemiologic investigation," he said.

This attitude of scientific caution is not always welcomed by those who have lost a loved one to degenerative illness. "I would never misrepresent the reality that there are people who are terribly unhappy with us," Hirsch acknowledged. "I sometimes have to tell the family, 'You know what's in your heart. I have to go with the evidence.'" As difficult as such conversations are, he believes that truth ultimately allows the living to come to terms with the dead. In a lecture that he gave in 1997, Hirsch discussed the medical examiner's "sacred responsibility" to talk to the bereaved. "None of us lives comfortably with uncertainty," he observed. "Once they know and understand the facts, even if unpleasant, most people have the ability to make their peace with reality."

ONE SUNDAY IN EARLY MARCH, I drove down to Little Egg Harbor, two hours south of New York City, to see Joe and Linda Zadroga. It was a cold day, with a brisk wind that whipped sand up from the roadsides and drove it against the sides of saltbox houses and trailered boats. Anchoring the tip of a small cul-de-sac, the Zadroga house, with an observatory window and wide, steeply raked front steps, is so much larger than its neighbors that at first I mistook it for a church. In the driveway, a pair of vintage cars sat covered next to a silver Ford pickup whose rear window had been transformed into a memorial: the World Trade Center towers framed by an American flag, and the legend "James Zadroga, Detective NYPD. Gone but not Forgotten. Fidelis ad Mortem."

Joe greeted me at the door. A former cop, like his son—he was the chief of police in North Arlington—he is stocky, with a bullish neck and shoulders and expressive hooded eyes. His wife, Linda, a petite woman with a deep tan, seemed tiny next to him. She wore a bright-pink sweatsuit and a silver heart-shaped pendant with a photograph of her son, inscribed "Jimmy." Sitting in the breakfast room, which overlooks a stretch of silvery shoreline, Joe talked about nearby Atlantic City ("She goes sometimes. I don't"), and bowed his head when Linda groused about his attachment to old T-shirts ("He holds on to everything"). Tyler Ann wandered in—a wary six-year-old in pink camouflage pajamas—and fussed with Joe's chair before wandering out again. When the conversation turned to James, Linda excused herself. "She don't like to talk about it," Joe said with a shrug.

Joe was just back from Washington, D.C., where he had attended a rally in support of the James Zadroga 9/11 Health and Compensation Act, introduced by Representative Maloney. Since his son's death, Joe has become a regular presence at such events; last year he attended about one a month. The attention that followed Breton's autopsy has added a disorienting public overlay to an otherwise private sorrow, and Joe seemed distinctly conscious of the symbolic pressure. But he was less concerned with arguing the merits of 9/11-related compensation claims than with affectionately recalling his son's good-old-boy ways. He cheerfully described James as "a partyer," who, during his days as a beat cop, would cadge meals from the Fire Department—"He always knew which firehouses had the best cooks"—and who, assigned to work on street crime, had an intuitive ability to sense who in a crowd was concealing a gun. There was a good-natured quality to this banter: a cop's-cop fondness, couched in jokey bravado. "That's his tough guy," Joe said, sifting through photographs of James that included one of him posing solemnly with a foot-long striper. He lingered over a later shot, taken at the ceremony when James made detective. In the picture, James stands with one arm around his wife, a pretty brunette with a shy smile. Dark-haired and baby-faced, he is powerfully built—a younger version of his father. He regards the camera with a lightly mocking pride.

On the morning of 9/11, James Zadroga had gone into the city early for an arraignment, and was almost back home before he switched on the radio. When he heard the news, he dashed inside, grabbed a bag, and turned around. "His wife was on her knees crying for him not to go back," Joe told me. Ronda Zadroga was seven months pregnant at the time. Assigned to a traffic post, Zadroga instead made his way to the towers, where he joined the search for survivors.

In all, Zadroga spent roughly three weeks on the pile, but according to Joe his decline began almost immediately. "He was sick, like, the first day," he told me. "And he said he felt terrible from all the chemicals he inhaled. He felt like he had the flu." Joe urged his son to see a specialist, and eventually got him an appointment with a pulmonologist at the Columbia University Medical Center. "He examined Jimmy and did some tests and said come back in three, four weeks. And then, when we came back, he took us into his office and he says, 'I'm not going to treat you.' And I said, 'Why?' And he said, 'Nah. I can't treat you.' And he just walked out the door." The Columbia University Medical Center could not comment, owing to patient confidentiality, but Joe claimed that the same problem occurred with one doctor after another. "Anytime he went to a hospital after that, they would say, 'Oh, you're a hero! We'll take good care of you,'" he said. "And within two days they'd discharge him, and not do anything. You know, shoot him up with steroids, maybe, so he could breathe easier, something like that."

The steroid shots, Joe claimed, gave James episodes of 'roid rage, and he suspects that they may also have led to the degeneration of two disks in his back. Joe said that James was hospitalized in 2002 and prescribed a course of intravenous antibiotics. While in the hospital, he was given a shunt—a small plastic tube that allows nurses to administer medication intravenously—which was left in when Zadroga was discharged.

In January of 2003, Joe persuaded a doctor at the Deborah Heart and Lung Center, in New Jersey, to take a biopsy of James's lungs. The procedure removed three inch-long chunks of tissue, one from each lobe of Zadroga's right lung, and James had to spend about a week in the hospital recuperating. I contacted Deborah to verify the results of the biopsy, but the hospital, citing confidentiality rules, would not release them. Joe, however, assured me that the test showed the presence of substantial damage. "His lungs were, like, eighty per cent black," he said. It seemed strange, I told Joe, that no doctor would offer a diagnosis, or order additional tests, even after a biopsy showing such extensive damage. (If there's anything doctors like, it's ordering additional tests.) When I asked Joe why he thought so many doctors would have denied treatment, he blamed what he considered to be political pressure from the city, intended to discourage institutions from treating a potentially controversial patient. He also told me that the Police Department suspected Zadroga of malingering and often sent cars, and even a helicopter, to watch the house. (When I contacted the N.Y.P.D. about this, a spokesman denied surveillance or use of a helicopter, but said that the N.Y.P.D.'s Medical Division did make unannounced visits to Zadroga's home and that this is a standard procedure for an officer on long-term sick leave.)

By 2003, Zadroga's condition had worsened further. He relied on oxygen tanks, and on three occasions passed out while driving, twice wrecking the car. Zadroga had already petitioned the Police Department for permanent retirement with a disability pension—a request that was granted in 2004. In the meantime, he and Ronda relocated to Florida, hoping that the climate would be easier on James's lungs. They bought a house not far from Ronda's parents, who sometimes babysat Tyler Ann, who was then a year old.

In Florida, however, events took a strange turn. About a year after the move, Ronda died, suddenly, at the age of twenty-nine. The news came as a surprise to Joe. "We didn't even know she was sick!" he told me. When I asked what could have killed Ronda so abruptly, Joe suggested that it was stress—the burden of looking after a young child and an invalid husband. "It was just too much for her," he said. Reached by phone, Ronda's mother would not talk about her daughter's death. A family friend, however, said that the actual cause was blood poisoning, brought on by intravenous drug use, adding that "everybody down here knew what was going on." A local investigation noted that, according to the Medical Examiner's Office, Ronda's lethal infection could not be conclusively determined to be the result of intravenous drug use. But track marks and multiple needle punctures were found on the body, and a toxicology report revealed nontoxic levels of drugs, including methadone.

There were other indications that something was seriously wrong. Arriving at the house in Florida for the funeral, Joe and Linda found it a mess. "Jimmy had passed out," Joe said. "The dogs had crapped everywhere." He cleaned up, and the next day a woman from the Florida Department of Children and Families arrived, escorted by two sheriffs. In a report filed with the DeSoto County Sheriff's Office, investigators who went to the house two days after Ronda's death to look into a report of child endangerment and drug use observed that Tyler Ann "was verbal and appeared happy," but noted the presence of a pill bottle prescribed in a relative's name, and carpet stains "where the family pets had left feces." Joe said that he was also quizzed about Tyler Ann's care. "She said, 'What's going to happen to the baby?'" Joe recalled. "And I said, 'What do you mean, what's going to happen to the baby? The baby's going to be with her father.' And she said, 'Well, her father can't take care of the baby, I understand.' And I said, 'Yeah, you understand right. He can't take care of the baby, but that's why he's going back to New Jersey with us.'"

Relocated to the house in Little Egg, James spent most of his time in bed. There, Joe remembered, he would sometimes sleep without waking for several days—"not to eat, not to go to the bathroom, nothing." By that point, Joe said, James was taking up to twenty pills a day for pain, including the drug OxyContin. As he told me this, Joe glanced out the window. "I never seen him use his drugs inappropriately," he said. "I can say that." He stopped, and pressed the heel of both hands to his eyes. "Whether he did . . ." He trailed off and sighed. "And then, on the other hand, like the lawyer said, what difference does it make anyway? He's dying anyway. That's not what killed him!"

LAST OCTOBER, SHORTLY AFTER RECEIVING a letter from the Office of the Chief Medical Examiner challenging the cause of James's death, the Zadroga family drove to New York to meet with Hirsch and hear his findings in detail, an experience that Joe described as "like getting hit over the head with a baseball bat." Joe recalled his wife's outrage. "She said, 'Are you calling my son a junkie? A drug user?' He said, 'No, no. I'm not. I'm not saying he used illegal drugs. They were legal drugs, but he used them wrong.'" The day after the meeting, Representative Maloney called. According to Joe, she gave her assurance that her opinion was unchanged, and urged him to call a press conference. "She said, 'We don't believe it. We'll fight it,'" he recalls.

Since then, the Zadroga story has been at a stalemate, with many officials unwilling to question either the accuracy of Hirsch's findings or the truthfulness of the Zadrogas' account. Shortly after the news broke, Mayor Bloomberg aroused fury among victims' families and in the press when he said, "We wanted to have a hero, and there are plenty of heroes. It's just, in this case, science says this was not a hero." He soon apologized for his remark but stopped short of retracting it. The Mayor called Zadroga "a great N.Y.P.D. officer," adding opaquely, "It's a question of how you want to define what a hero is, and certainly I did not mean to hurt the family or impugn his reputation."

The question of what actually killed James Zadroga is one that public officials appear reluctant to answer candidly, and, increasingly, the debate surrounding the case seems to be less about scientific evidence than about public feeling. This is the view of Brian Monahan, a sociologist at Iowa State University who is working on a book on 9/11. "After 9/11, we suddenly had emerging all these perfect heroes, who stood against bin Laden, the perfect villain," he said, observing that Zadroga was also a natural icon: young, the first to die, a father. Aligning oneself with a hero, he added, is a convenient way to manage the chaos of a complicated event, and to present a point of view that can't be argued with. In that sense, he said, Zadroga was a valuable figure for politicians, 9/11 groups, and police unions—"anyone with a vested interest in how 9/11 illness was going to be treated."

When I spoke with Representative Maloney, she acknowledged the "confusion" created by Hirsch's finding. But she defended the idea that Zadroga's time at Ground Zero ultimately precipitated his decline, and the bill she introduced, the James Zadroga 9/11 Health and Compensation Act, is still making its way through Congress. Regardless of what eventually happened, Maloney argued, "James Zadroga would still be alive if not for 9/11."

The truth of this is hard to gauge. Four months after meeting with Joe, I wrote to the Zadrogas' lawyer, who agreed to fax me a copy of the results of James's 2003 biopsy, the one that Joe had said showed extensive lung damage. The document that arrived was eight pages long: a comprehensive report on the various tests and their findings. But its import was unequivocal. The lungs had only minor abnormalities and showed no evidence of talc or cellulose. When I contacted a pathologist unconnected with the case to ask whether this could have been an oversight, he scoffed, explaining that, under the polarized light that labs use to spot foreign particulate matter, such particles shine out like stars in the night sky. Had the material Breton found in Zadroga's lungs in 2006 been there in 2003, it would have lit up the lab.

A few days after receiving the biopsy report, I contacted Dr. Philip Landrigan, the physician who oversees Mt. Sinai's World Trade Center Medical Monitoring and Treatment Program. Landrigan has spent the past seven years tracking the health of twenty-five thousand workers and volunteers exposed to the dust cloud at Ground Zero and advocating additional funding for treatment and research. Zadroga's celebrity, he acknowledged, had helped advance the political cause of 9/11 survivors by "putting a human face on the statistics." When I asked whether this development might now create problems—whether there could be a backlash if it became known that Zadroga's death may not have been caused by exposure at Ground Zero—Landrigan said that he hoped it wouldn't. As for Zadroga's continued fame as the first dust victim, he shrugged it off. "It is what it is," he said. "It's a fact at this point."

But, while the biopsy sheds some light on the facts of Zadroga's death, it seems to make the events of his life even murkier. If Zadroga's lungs weren't significantly compromised, why did he undergo a biopsy? And if he wasn't mainlining drugs in early 2003—as the biopsy would indicate—when, and why, did he start?

Before I left Little Egg, Joe gave me a letter that he said James had written on the first anniversary of the attacks, at the behest of Ronda's father, a minister. It was morbid and bleak. Recalling his time at the towers, Zadroga described a landscape strewn with human remains: an ear fringed with hair, a dismembered foot. A year later, he reported that his health was bad—"I can't breathe, my throat is constantly sore, I'm always coughing," he wrote—and that he remained deeply troubled by his days at Ground Zero, plagued by nightmares, sleeplessness, and anxiety. Zadroga's memories seemed to overshadow even the birth of his first child, and he apparently saw himself as doomed. "Everyone praises the dead as heroes, as they should, but there are more living suffering than dead," he wrote. "The dead, their deaths were quick and painless, and mine has just begun."

JOHN HORGAN

War! What Is It Good For? Absolutely Nothing

FROM *DISCOVER*

War is as old as civilization, if not older. But is it unavoidable? John Horgan finds some researchers who argue that war is "evitable."

FRANS DE WAAL STANDS IN A WATCHTOWER AT THE Yerkes National Primate Research Center north of Atlanta, talking about war. As three hulking male chimpanzees and a dozen females loll below him, the renowned primatologist rejects the idea that war stems from "some sort of blind aggressive drive." Observations of lethal fighting among chimpanzees, our close genetic relatives, have persuaded many people that war has deep biological roots. But de Waal says that primates, and especially humans, are "very calculating" and will abandon aggressive strategies that no longer serve their interests. "War is evitable," de Waal says, "if conditions are such that the costs of making war are higher than the benefits."

War evitable? That is a minority opinion in these troubled times. For several years I've been probing people's views about war. Almost everyone, regardless of profession, political persuasion, or age, gives me the same answer: War will never end. I asked 205 students at the college where I teach, "Will humans ever stop fighting wars, once and for all?" More than 90 percent said no. This pessimism seems to be on the rise; in the mid-1980s, only one in three students at Wesleyan University agreed that "wars are inevitable because human beings are naturally aggressive."

Asked to explain their views, most fatalists offer variations on Robert McNamara's remarks in the documentary *The Fog of War.* "I'm not so naive or simplistic to believe we can eliminate war," said McNamara, who was the U.S. defense secretary during the Vietnam War. "We're not going to change human nature any time soon." War, in other words, is inevitable because it is innate, "in our genes," as my students like to put it.

This dark outlook seems confirmed not only by the daily barrage of headlines from war-torn regions around the world—Iraq, Afghanistan, Congo—and the seemingly endless threat of terrorism, but also by findings from primatology, anthropology, and other fields. Over the last few decades, researchers in Africa have observed males in rival troops of chimpanzees raiding and killing each other. Archaeologists and anthropologists also keep unearthing evidence of warfare in their studies of prehistoric and tribal human societies.

De Waal acknowledges that "we have a tendency, and all the primates have a tendency, to be hostile to non-group members." But he and other experts insist that humans and their primate cousins are much less bellicose than the public has come to believe. Studies of monkeys, apes, and *Homo sapiens* offer ample hope that we can overcome our aggressive tendencies and greatly reduce or maybe even eliminate warfare.

Biologist Robert Sapolsky is a leading challenger of what he calls the "urban myth of inevitable aggression." At his Stanford University office, peering out from a tangle of gray-flecked hair and beard, he tells me that primate studies contradict simple biological theories of male belligerence—for example, those that blame the hormone testosterone. Aggression in primates may actually be the *cause* of elevated testosterone, rather than vice versa. Moreover, artificially increasing or decreasing testosterone levels within the normal range usually just reinforces previous patterns of aggression rather than dramatically transforming behavior; beta males may still be milquetoasts, and alphas still bullies. "Social conditioning can more than make up for the hormone," Sapolsky says.

Environmental conditions can also override biology among baboons, who, much like chimpanzees, seem hardwired for aggression. Since early 1978, Sapolsky has traveled to Kenya to spy on baboons, including Forest Troop, a group living near a tourist lodge's garbage dump. Because they had to fight baboons from another troop over the scraps of food, only the toughest males of Forest Troop frequented the dump. In the mid-1980s, all these males died after contracting tuberculosis from contaminated meat.

The epidemic left Forest Troop with many more females than males, and the remaining males were far less pugnacious. Conflict within the troop dropped dramatically; Sapolsky even observed adult males grooming each other. This, he points out in an article in *Foreign Affairs*, is "nearly as unprecedented as baboons sprouting wings." The sea change has persisted through the present, as male adolescents who join the troop adapt to its mores. "Is a world of peacefully coexisting human Forest Troops possible?" Sapolsky asks. "Anyone who says, 'No, it is beyond our nature,' knows too little about primates, including ourselves."

Sapolsky is hardly a starry-eyed optimist. He doubts whether acts of large-scale violence will ever completely vanish. Yes, the threat of war between major powers has declined, he notes, but the ability of small groups or even individuals to wreak enormous havoc—with nuclear, chemical, or biological weapons, not to mention jumbo jets—has grown. "So at a certain level the danger has risen, if not the sheer incidence," he says. Nonetheless, Sapolsky believes that "there is a great potential for dramatically decreasing the frequency of war and getting a lot better at intervention, termination, and reconciliation."

De Waal, who met me at the Yerkes center after attending a disarmament workshop in Geneva, agrees that aggression is part of our nature. So too, he adds, are cooperation, conflict resolution, and reconciliation. For decades he has carefully documented how apes and monkeys avoid fights or quickly make up after them by sharing food, grooming each other, or even hugging and kissing.

These traits are especially pronounced in the ape species *Pan paniscus*. More commonly known as bonobos, they are darker-skinned and more slender than common chimpanzees and have markedly different lifestyles. "No deadly warfare," de Waal says, "little hunting, no male dominance, and enormous amounts of sex." Their promiscuity, he speculates, reduces violence both within and between bonobo troops, just as intermarriage does between human tribes. What may start out as a confrontation between two bonobo communities can turn into socializing, with sex between members, grooming, and play.

De Waal suspects that environmental factors contribute to the bonobos' benign character; food is more abundant in their dense forest habitat than in the semi-open woodlands where chimpanzees live. Indeed, his experiments on captive primates have established the power of environmental factors. In one experiment, rhesus monkeys, which are ordinarily incorrigibly aggressive, grew up to be kinder and gentler when raised with mild-mannered stump-tailed monkeys.

De Waal has also reduced conflict among monkeys by increasing their interdependence and ensuring equal access to food. Applying these lessons to humans, de Waal sees promise in alliances, such as the European Union, that promote trade and travel and hence interdependence. "Foster economic ties," he says, "and the reason for warfare, which is usually resources, will probably dissipate."

The question that divides primate researchers—whether war is innate in us and in our hairier kin—has also challenged anthropologist Douglas Fry, whose interest goes back to his teenage years, when the Vietnam War was still raging. He recalls wondering, "Is this something we always have to live with, war after war after war?" His research, says Fry, who left the United States in 1995 to accept a position at Åbo Akademi University in Finland, has led him to reject this conclusion. "Warfare is not inevitable," he insists in his book *Beyond War*, because humans "have a substantial capacity for dealing with conflicts nonviolently."

Fry notes that the earliest widely accepted evidence of possible warfare is a mass grave of skeletons with smashed skulls and hack marks found near the Nile River; the grave dates back some 12,000 to 14,000 years. Such evidence accumulates from later periods as humans around the world abandoned a nomadic existence for a more settled one, leading eventually to the creation of agriculture and states. This evidence consists not only of mass graves but also of weapons clearly designed for fighting, fortified settlements, and rock art depicting battles.

FRY HAS ALSO IDENTIFIED 74 "nonwarring cultures" that—while only a fraction of all known societies—nonetheless contradict the depiction of war as universal. His list includes nomadic hunter-gatherers such as the !Kung in Africa and Aborigines in Australia. These examples are crucial, Fry says, because our ancestors are thought to have lived as nomadic hunter-gatherers from the emergence of the Homo lineage just over 2 million years ago in Africa until the appearance of agriculture and permanent settlements about 12,000 years ago. That time span constitutes 99 percent of our history.

Lethal violence certainly occurred among those nomadic hunter-gatherers, Fry acknowledges, but for the most part it consisted not of genuine warfare but of fights between two men, often over a woman. These fights would sometimes precipitate feuds between friends and relatives of the initial antagonists, but members of the band had ways to avoid these feuds or cut them short. For example, Fry says, third parties might step between the rivals and say, "'Let's talk this out' or 'You guys wrestle, and the winner gets the woman.'"

Fry has sought to determine what distinguishes peaceful societies from more violent ones. One clue comes from his fieldwork among the Zapotec, peasant farmers descended from an ancient, warring civilization in Oaxaca, Mexico. There, Fry studied two Zapotec communities, which he labeled with the pseudonyms San Andreas and La Paz. San Andreas's rates of male-on-male violence, spousal abuse, and child abuse are five times higher than those in La Paz. The reason, Fry suspects, is that women in La Paz have long contributed to the income of their families by making and selling pottery, thus earning the respect of the males.

Fry believes that empowering females may reduce the rate of violence committed within and by a nation. He notes that in Finland—which has a low rate of crime and violence compared with other developed countries—a majority of the cabinet ministers and more than 40 percent of the members of Parliament are women. "I don't see this as a panacea," Fry adds, recalling "iron lady" Margaret Thatcher, "but there are good reasons for having a balance of the more caring sex in government."

The anthropologist Richard Wrangham is one of several scientists at Harvard who present a much darker view of human nature than Fry does. In his 1996 book, *Demonic Males: Apes and the Origins of Human Violence* (co-authored with Dale Peterson), Wrangham argues that "chimpanzee-like violence preceded and paved the way for human war, making modern humans the dazed survivors of a continuous, 5-million-year habit of lethal aggression." Natural selection has favored combative, power-hungry males, he contends, "because with extraordinary power males can achieve extraordinary reproduction."

"I worked in the Congo," Wrangham remarks drily when I call him in England, where he is en route to Africa to study chimpanzees. "It's hard for me to feel that we're a peaceful species when you have hundreds of thousands of people being killed there." Wrangham says de Waal is exaggerating the significance of the bonobos, and he scoffs at Fry's attempt to minimize warfare among hunter-gatherers by excluding "feuds."

But like his more sanguine colleagues, Wrangham believes we can overcome our propensity for aggression. Primate violence is not blind and compulsive, he asserts, but rather calculating and responsive to circumstance. Chimpanzees fight "when they think they can get away with it," he says, "but they don't when they can't. And that's the lesson that I draw for humans." Wrangham notes that male hunter-gatherers within the same band rarely kill each other; their high mortality rates result from conflict between groups.

Wrangham even agrees with Fry on how to decrease conflicts both between and within nations. He points out that as female education and economic opportunities rise, birthrates tend to fall. A stabilized population lessens demands on governmental and medical services and on natural resources; hence, the likelihood of social unrest also decreases. Ideally, Wrangham says, these trends will propel more women into government. "My little dream," he confesses, is that all nations give equal decision-making power to two entities, "a House of Men and a House of Women."

Like Wrangham, the archaeologist Steven LeBlanc is critical of scientists who emphasize the peaceful aspects of human nature. At Harvard's Peabody Museum of Archaeology and Ethnology, where he serves as director of collections, LeBlanc points to a piece of carved wood hanging on his office wall. This, he notes, is a spear employed by Australian Aborigines (who, according to Fry, rarely or never waged war). A short, bearded, excitable man, LeBlanc accuses Fry of perpetuating "fairy tales" about levels of violence among hunter-gatherers and other pre-state people.

* * *

LEBLANC CONTENDS THAT RESEARCHERS HAVE unearthed evidence of warfare as far back as they have looked in human prehistory, and ethnographers have observed significant levels of violence among hunter-gatherers such as the !Kung. In his book *Constant Battles: Why We Fight* (with Katherine E. Register), he espouses a bleak, Malthusian view of human prehistory, in which war keeps breaking out as surging populations outstrip food supplies. Warfare, he writes, "has been the inevitable consequence of our ecological-demographic propensities."

Still, when asked point-blank if humans can stop fighting wars, LeBlanc replies, "Yes, I think it's completely possible." He notes that many warlike societies—notably Nazi Germany and imperial Japan and even the Yanomami, a notoriously fierce Amazonian tribe—have embraced peace. "Under certain circumstances," he says, warfare "stops on a dime" as a result of ecological or cultural change. Two keys to peace, he believes, are controlling population growth and finding cheap alternatives to fossil fuels. "I was just in Germany," LeBlanc exults, "and there are windmills everywhere!"

Despite the signs of progress against our belligerent side, all these scientists emphasize that if war is not inevitable, neither is peace. Major obstacles include religious fundamentalism, which not only triggers conflicts but also contributes to the suppression of women; global warming, which might produce ecological crises that spur social unrest and violence; overpopulation, particularly when it produces a surplus of unmarried, unemployed young men; and the proliferation of weapons of mass destruction.

Moreover, all the solutions to war come with caveats. Sapolsky suggests that eliminating poverty, while an important goal in its own right, may not extinguish war in all regions. Among baboons, lions, and other animals, aggression sometimes "goes up during periods of plenty because you have the energy to waste on stupid stuff rather than just trying to figure out where your next meal is coming from." De Waal raises concern about female empowerment. Studies of apes and humans, he says, have found that while females fight less frequently than males, when they do fight, they "hold grudges much longer."

A crucial first step toward ending war is to reject fatalism, in ourselves and in our political leaders. That is the view of the Harvard biologist Edward O. Wilson, who is renowned for his conservation efforts as well as for his emphasis on the genetic underpinnings of social behavior. A rangy man with a raptor's long, narrow nose and sharp-eyed gaze, Wilson has not budged from his long-standing position that the propensity for group aggression, including war, is deeply ingrained in our history and nature. He notes, however, that group aggression is highly "labile," taking many different forms and even vanishing under certain circumstances.

He is therefore confident that we will find ways to cease making war on nature as well as on each other, but it is a race against time and human destructiveness. "I'm optimistic about saving a large part of biodiversity," he says, "but how much depends on what we do right now. And I think that once we face the problems underlying the origins of tribalism and religious extremism—face them frankly and look for the roots—then we'll find a solution to those, too, in terms of an informed international negotiation system." Wilson pauses and adds, "We have no option but optimism."

MARINA CORDS

Face-Offs of the Female Kind

FROM *NATURAL HISTORY*

Like most primates, African blue monkeys fight each other; what may be surprising, however, is that among blue monkeys the females do the fighting. A longtime researcher of blue monkeys, Marina Cords reports on this behavior and comes upon an evolutionary puzzle.

URGENT, LOW-PITCHED GROWLS EMANATE FROM DENSE foliage in the Kakamega Forest in western Kenya. The growls are punctuated by birdlike chirps and, every so often, a deep, resonant boom. The sounds get louder as two groups of blue monkeys, which are gray and black rather than blue, face off across an invisible line in the tangle of branches. Each group is about forty strong. In the front lines are adult females, the size of house cats, literally shoulder to shoulder, glaring and threatening. The opposing phalanxes move in parallel, sometimes pushing forward, sometimes retreating. Individual "soldiers" join and leave the front ranks; behind them, others—adult females, the juveniles, and the group's one resident male—chirp and growl in support, or simply sit and watch. Still others, usually at a distance from the commotion, seem completely unconcerned and uninvolved in the battle.

In my fieldwork on blue monkeys (*Cercopithecus mitis*), my colleagues and I have witnessed fierce encounters of that kind every month or so, while less dramatic confrontations, such as a couple of animals growling at their neighbors, occur every day or two. What's in dispute is territory, territory that contains the fruits, insects, flowers, and leaves that make up most of the diet of these monkeys. The battles are remarkable both for the ganglike cohesiveness they bring to what can otherwise seem a scattered, unconnected group, and for the intensity of the aggression that can erupt among females with normally peaceful demeanors. For it is the females that participate in these turf wars, grabbing and biting their opponents when aggressive threats are not enough.

Chimpanzees, our close relatives, are sometimes compared to politicians: they engage in power plays; they use diplomacy; they assign perks to various positions in a complicated social hierarchy. Blue monkeys, which seem calmer and less prone to form coalitions, appear to be more egalitarian. And yet my study of the group dynamics of territorial battles, and how these may figure into the way a group later splits into smaller new groups, has revealed unexpected complexity in blue monkey social structure. As in some human political organizations, those on top may depend more on those at the bottom than first meets the eye.

THE FEEDING TERRITORIES OVER WHICH blue monkeys battle are often so specific that a person could draw lines to demarcate them: this tree belongs to this group, the next tree over doesn't.

Groups can coexist peacefully very near one another as long as each stays on its side of these imaginary lines.

Territorial behavior occurs in many animals that defend feeding grounds, breeding areas, and in a few cases, courtship arenas. Among primates, territorial defense is just one way a group cooperates to compete with the neighbors. Not all primates are territorial in the way blue monkeys are, defending a specific piece of land. However, many primate groups react aggressively toward neighboring groups wherever they meet. Biologists would not call such populations territorial, but there is nevertheless cooperative defense of resources.

Quite a few researchers have investigated what certain primate species fight over; less often have they looked at how group members differ in their participation. Of those studies, almost all involve species in which the males do most of the fighting. My study focused on the question of why certain individual females did the fighting.

At first glance, it seems unfair that only some members of a blue monkey group would fight to defend a food supply that benefits all the group members—an effort that economists call "collective action for a public good." Indeed, those monkeys that sit out the battles are getting a free ride. As a matter of fact, any individual monkey would seem well advised to avoid joining an intergroup battle: it's a way to get the benefit (more food) without paying any cost (potential injury or even death).

The most severe wounds I've seen a female sustain in my twenty-nine years of study occurred during a group battle. For more than an hour, members of the front lines had taken each other on, sometimes grappling fiercely in a tangle as a roar of growls, chirps, and loud screams erupted. When the losing group finally retreated, I moved forward to get a closer look. Normally monkeys are wary of people, so I was amazed that I could come within six feet of an adult female. She was breathing heavily as she lay prostrate on some low branches, just level with my eyes. Her body was covered with flecks of blood—she must have been bitten all over, severely and repeatedly. Only after lying perfectly still for a half hour did she pull herself together and slowly move off to a safer distance.

More recently, in another prolonged, loud, and vicious group fight, another female had the skin ripped off the back of her calf, starting from the knee. The skin bunched down at her ankle like a frilly bobby sock, leaving the red calf muscle fully exposed; eventually she limped away, though carefully and in obvious pain. Both of those animals survived their injuries, but that was a matter of chance: open wounds in a tropical environment can become infected and lead to death.

And yet, if every monkey chose to take a free ride to avoid such risks, no cooperative territorial defense would occur, and all would lose out. This dilemma is what economists refer to as a "collective action problem." Evolutionary biologists think in a parallel manner, because natural selection should act on animal decision making in a way that maximizes individual fitness. In other words, natural selection should favor selfish and self-protective decision making, which maximizes an individual's chances to survive and reproduce. If taking a free ride offers the best deal to the individual, collective action should always fall apart.

To understand from an evolutionary perspective why some members of a group of blue monkeys fight even though others ride free, it seems that there must be additional important factors that a simple model of cooperation does not take into account. For one thing, individual group members are not identical: one might expect some difference in participation if the costs and benefits of territorial defense are not the same for all potential participants.

Over a five-year period, my colleagues and I monitored individuals' participation in territorial encounters in five study groups of the blue monkey subspecies *C.m. stuhlmanni* in the Kakamega Forest to see if such disparities in individuals' costs and benefits offered a compelling explanation for divergences in participation. As noted before, adult females regularly joined the fray, while adult males engaged infrequently—and when they did, it was mostly to vocalize from the back lines. (Although each blue monkey group usually contains only one resident adult male, an influx of male adults sometimes swells the group during the three- to five-month breeding season.) Juveniles were less likely than adults to get involved in the intergroup aggression; generally, however, the older the juveniles were, the more likely they were to join in. At every age, juvenile females participated more than juvenile males—foreshadowing the adult pattern.

We can in fact begin to explain these results in terms of the relative costs and benefits of participating in intergroup battles. From an evolutionary perspective, the relative reproductive success of female mammals is limited mainly by access to food: enduring a pregnancy, and in particular, lactating to support a growing infant, are energy-intensive life stages limited to adult females. Thus females have especially much to gain from defending a source of food, and it makes sense that they are more involved than males in defending feeding territories.

In contrast to females, the relative reproductive success of male mammals depends mainly on their access to female mates, at least in cases where paternal care is minimal (as is the case in blue monkeys). When male primates participate in intergroup aggression, they are usually defending mates, not food. Some exceptions to this generalization have recently been discovered: Eastern black-and-white colobus monkey males seem to defend feeding areas directly, for example. But researchers believe this may actually be a way to attract potential mates, rather than a defense of feeding grounds for their own sake. In blue monkeys, however, males are obviously less involved in battles between groups than females are, and it seems clear that defending food sources is not one of their strategies for attracting mates.

The age-related difference in participation, by contrast, may reflect the differing risk to individuals. Given the potential for extreme violence, smaller and more vulnerable individuals may be less inclined to take a risk. Not only juvenile monkeys, but also females with infants, are less likely to take part in intergroup aggression. This observation is also consistent with the idea that individuals make decisions based on their personal risks: a female carrying an infant is probably less agile in the branches, and agility is at a premium in the fast-moving action of escalated encounters.

Why, then, do females of higher rank in the dominance hierarchy participate more in territorial battles than do those of lower rank? Rank structure in blue monkeys is linear: if A is dominant to B and B is dominant to C, then A is dominant to C as well. In that respect blue monkey hierarchies resemble those of many other Old World monkeys, such as the rhesus macaque and the baboon. What's different is that blue monkey rank structure is less "in your face": dominance interactions are rare, especially those more aggressive ones in which there is a clear loser. Instead, dominance is revealed over the long term by how a lower-ranking individual will tend to avoid a higher-ranking one or withdraw in her favor. It's a muted dance of approach and retreat in which the dancers perform occasionally. In fact, only after piecing together years of observations to detect repeated patterns did we perceive a hierarchy in blue monkeys at all.

Rank is generally a poor predictor of female blue monkey behavior, so we had not expected dominance rank to play a role during territorial battles. For example, females of lower rank do not travel more or farther to get food, do not spend more time feeding, and do not have a lower-quality diet than those of higher rank. Nor do they occupy spatial positions in trees (at the edges of a tree's crown) or in the group as a whole (on the edge of the group) that would expose them more to predators. Most important, there is no evidence that they breed less successfully: neither the rate at which they produce offspring nor the survival of their infants suffers compared with that of higher-ranking females.

If high-ranking females possessed an advantage in any of those ways, one might understand their greater participation in group battles: they are the animals with the most to gain. But for blue monkeys there must be a more subtle explanation.

I can suggest two possibilities. The first is that participation in cooperative territorial defense is traded for other social services. Here behavioral biologists use another economics analogy: the idea that cooperative exchange can involve multiple "currencies." At the simplest level, one animal takes on the risks of an intergroup battle, while another pays back with a different social good. In some primate species, teaming up in aggressive alliances to sort out squabbles within a group is a significant service provided by friends, but in blue monkeys this seems too rare to be important. Other obvious contenders are social grooming, or perhaps tolerance while feeding: a monkey is allowed to feed undisturbed, rather than be chased away by a higher-ranking group member. And, for reasons that are as yet unclear, high- and low-ranking animals "play" with different currencies. At present we don't know whether such exchanges occur, but our team is monitoring exchange in all three currencies (territorial fighting, grooming, and feeding tolerance) to find out.

When two animals exchange services with each other, reciprocity is direct: I give something to you, and you pay me back in some way. But reciprocity can also be indirect, when an individual provides services to another and is paid back (whether in the same currency or not) by a third party or parties. In theoretical models, indirect reciprocity can persist when individuals have reputations, and when their fate in a marketplace of social partners depends on their reputations. Both theoretical models and experiments with human players suggest that investment in reputation occurs. For example, human subjects will donate to a public good if doing so enhances their reputation, and if others can use information about their reputation to make decisions about cooperating with them in the future. So, contribution to the public good can eventually bring benefits to the contributor, lessening the temptation to ride free. Whether nonhuman primates attend to reputations or practice indirect reciprocity is, however, currently unknown.

The second potential explanation for why high-ranking females fight invokes the theory of "reproductive skew." This theory was originally devised to explain how reproduction is distributed among the members of a society. One might ask, for example, why high-ranking individuals do not always prevent lower-ranking ones from reproducing. A possible answer to this question is that lower-ranking individuals have something the members of higher rank need. For example, perhaps the social unit needs all of its members alive and well for any of them to reproduce successfully. If so, a higher-ranking individual might need to give up some control over reproduction to keep lower-ranking individuals around. Without such an incentive, the latter might go elsewhere.

That general idea may indeed apply to blue monkeys. Specifically, high-ranking females may need to keep low-ranking ones around to cope effectively with the aftermath of group fissions. Blue monkey groups tend to split when they reach fifty to seventy members. The reasons for such divisions are not well known in this species, but the consequences seem clear. The two new daughter groups, which are seldom the same size, engage in a series of intense intergroup battles to divvy up the original territory.

In fact, that is when territorial boundaries—normally stable over generations—change the most. Invariably, the smaller daughter group ends up with the smaller parcel of land, which may not be big enough for its needs. That new home range seldom includes the most visited portions of the original group's range, and it often includes areas that were relatively infrequently used.

The fact that there are different consequences for large and small daughter groups is one piece of the puzzle. Another is that lower-ranking females appear to have some control over how the group divides. They may choose not to join higher-ranking females in a new group, for example; in addition, they may actively and collectively exclude higher-ranking females. In one fission, we witnessed lower-ranking females harassing a higher-ranking matriarch that tried to join their subgroup. She was highly motivated to join it, as it included almost all of her many offspring. In the face of repeated, targeted threats and chases by lower-ranking females, however, she ended up with the smaller subgroup of females, all higher ranking than she was. She was thus at the bottom of the hierarchy in her new group, and separated from all of her family except her young daughter—and this position seemed to be imposed on her from below.

Presumably it is in the best interest of high-ranking females to join the larger subgroup during a group fission, but they risk being abandoned by lower-ranking individuals looking for a better situation elsewhere. If high-ranking individuals take on a greater share of territorial defense, they may create the staying incentive needed to retain a loyal following, thus ensuring the most advantageous outcome.

Our field research continues to investigate such questions of social structure by focusing on the behavior of individuals and their exchanges with one another, both one-on-one and in the context of group territoriality. This may prove to be a long undertaking, as some of the relevant social events occur rarely. Escalated attacks happen only once a month or so, and group fissions take place on average only once in ten years. With females living as long as thirty years or more, perhaps one should expect social lives that play out over a protracted period, but that presents scientists with a challenge. It's a good thing that human researchers live longer than their subjects!

MARTIN ENSERINK

Tough Lessons from Golden Rice

FROM *SCIENCE*

Ten years ago golden rice—rice bioengineered to fight vitamin A deficiency—was poised to save millions of people around the world. It hasn't worked out that way. Martin Enserink looks into the troubled history of a promising breakthrough.

IT'S EASY TO RECOGNIZE INGO POTRYKUS AT THE TRAIN station in Basel, Switzerland. Quietly waiting while hurried travelers zip by, he is holding, as he promised, the framed and slightly yellowed cover of the 31 July 2000 issue of *Time* magazine. It features Potrykus's bearded face flanked by some bright green stalks and a bold headline: "This Rice Could Save a Million Kids a Year."

The story ran at a time when Potrykus, a German plant biotechnologist who has long lived in Switzerland, was on a roll. In 1999, just as he was about to retire, Potrykus and his colleagues had stunned plant scientists and biotechnology opponents alike by creating a rice variety that produced a group of molecules called provitamin A in its seeds. The researchers thought this "golden rice"—named for the yellow hue imparted by the compounds—held a revolutionary promise to fight vitamin A deficiency, which blinds or kills thousands of children in developing countries every year.

Almost a decade later, golden rice is still just that: a promise. Well-organized opposition and a thicket of regulations on transgenic crops have prevented the plant from appearing on Asian farms within two to three years, as Potrykus and his colleagues once predicted. In fact, the first field trial of golden rice in Asia started only this month. Its potential to prevent the ravages of vitamin A deficiency has yet to be tested, and even by the most optimistic projections, no farmer will plant the rice before 2011.

The delays have made Potrykus, who lives in Magden, a small village in an idyllic valley near Basel, a frustrated man. For working on what he considers a philanthropic project, he has been ridiculed and vilified as an industry shill. Relating the golden rice saga at his dinner table while his wife serves croissants and strong coffee, he at times comes off as bitter.

There's more at stake than golden rice and personal vindication, he says. In his view, two decades of fear-mongering by organizations such as Greenpeace, his prime nemesis, have created a regulatory climate so burdensome that only big companies with deep pockets can afford to get any genetically modified (GM) product approved. As a result, it has become virtually impossible to use the technology in the service of the poor, Potrykus says.

Not everybody is so gloomy. Potrykus's co-inventor and main partner, plant biochemist Peter Beyer of the University of Freiburg in Germany, agrees that it's been a difficult decade. But a more cheerful character by nature, Beyer believes rules are just something to be dealt with; complaining about them does little, he says. A handful of other researchers working on GM crops to fight malnutrition also feel confident that their work will eventually pay off.

Many scientists agree with Potrykus, however, that GM technology has become so controversial that for now, there's little point in harnessing it for the world's poorest. HarvestPlus, a vast global program at public research institutes aimed at creating more nutritious staple crops, is forgoing GM technology almost entirely and using conventional breeding instead, despite its built-in limitations. GM products just might end up on the shelf, says HarvestPlus Director Howarth Bouis.

Potrykus, now 75 years old, worries that he may not live to see his invention do any good. "It's difficult for me not to get upset about this situation," he says.

A Dream Takes Root

The idea for golden rice was born at an international agricultural meeting in the Philippines in 1984, says Gary Toenniessen of the Rockefeller Foundation, a philanthropy in New York City. It was the early days of genetic engineering, and over beers at a guesthouse one evening, Toenniessen asked a group of plant breeders how the technology of copying and pasting genes might benefit rice. "Yellow endosperm," one of them said.

That odd answer alluded to the fact that a quarter-billion children have poor diets lacking in vitamin A. This deficiency can damage the retina and cornea and increase susceptibility to measles and other infectious diseases. The World Health Organization (WHO) estimates that between 250,000 and 500,000 children go blind every year as a result, and that half of those die within 12 months. Vegetables such as carrots and tomatoes, as well as meat, butter, and milk, can provide the vitamin or its precursors, but many families in poor countries don't have access to them. A rice variety producing precursors to vitamin A in its endosperm, the main tissue in seeds, might provide a solution—and it would have yellow kernels.

Classical breeding cannot produce such a rice, however, because although pro-vitamin A is present in the green parts of the rice plant, no known strain makes it in its seeds. The only option is to tinker with rice's DNA to produce the desired effect. Throughout the 1980s, the Rockefeller Foundation funded several exploratory studies, but the plan didn't gel until a brainstorming meeting in New York City in 1992, at which scientists discussed the bold idea of reintroducing the biochemical pathway leading to beta carotene, the most important pro-vitamin A, into rice but putting it under control of a promoter that's specific to endosperm.

Potrykus, then a pioneer in rice transgenics at the Swiss Federal Institute of Technology (ETH) in Zürich, attended, as did Beyer, who specialized in carotenoid biochemistry and molecular biology. The two met on the plane to New York and hit it off; their fields of expertise were complementary, and the fact that Zürich is less than two hours from Freiburg was helpful. They soon had a proposal written up.

Beyer admits he barely believed in the idea himself, and Rockefeller's scientific advisory board was equally skeptical. Introducing an entire genetic pathway into rice seemed like a stretch. Still, the foundation rolled the dice and supported the project.

It took seven years, but Potrykus and Beyer eventually succeeded in making golden rice by splicing two daffodil genes and a bacterial gene into the rice genome. The eureka moment arrived late one night in Freiburg, Beyer recalls. He was analyzing the molecular content of seeds produced in Potrykus's lab, as he often did, using a technique called high-performance liquid chromatography. This time, peaks showed up on the screen where they had never appeared before—the signals of carotenoids. When Beyer went back to look at the batch of seeds, he noticed something he had missed: The grains had a faint yellow hue. Golden rice had been born.

The Battle Begins

Potrykus says he always knew golden rice—a Thai businessman suggested the catchy name—would be controversial. As a professor in Switzerland, one of the most fiercely anti-GM countries in Europe, he had been confronted with angry students since the 1980s. To protect his plants, ETH spent several million dollars on a grenade-proof greenhouse. For Beyer, unofficial road signs declaring the Upper Rhine Valley a "GM technology–free region" are a twice-daily reminder that the climate in Germany isn't much better.

But golden rice posed a special dilemma to GM crop opponents, admits Benedikt Haerlin, who coordinated Greenpeace's European campaign at the time and now works for the Foundation on Future Farming. Unlike the existing GM crops that primarily helped farmers and pesticide companies, it was the first crop designed to help poor consumers in developing countries. It might save lives. The decision whether to oppose it weighed heavily on him, Haerlin says, which is why he consulted with WHO experts on vitamin A and why he traveled to Zürich to spend a day at Potrykus's lab to talk. Potrykus, impressed by Haerlin's intelligence, hoped to convince his fellow countryman.

He failed. Although Greenpeace pledged not to sabotage field trials, it did launch an aggressive campaign against golden rice. It argued that the crop was an industry PR ploy—seed company Syngenta was involved in the project, the group pointed out—designed to win over a skeptical public and open the door to other GM crops. Golden rice did not attack the underlying problem of poverty, Greenpeace said; besides, other, better solutions to vitamin A deficiency existed.

Perhaps Greenpeace's most effective argument, however, was that golden rice simply wouldn't work. The most successful strain created in 2000 produced 1.6 micrograms of pro-vitamin A per gram of rice. At that rate, an average two-year-old would need to eat 3 kilos of golden rice a day to reach the recommended daily intake, Greenpeace said, and a breastfeeding mother more than 6 kilos. To drive the point home, an activist in the Philippines sat down behind a giant mound of golden rice during a press conference. "Fool's gold," Greenpeace called it.

A photo of the event, which quickly found its way around the world, still makes Haerlin chuckle—and it still makes Potrykus angry. Greenpeace assumed that children had to get *all* of their vitamin A from rice, which was unrealistic; it also ignored the fact, says Potrykus, that even half the recommended intake may prevent malnutrition. And Greenpeace assumed that the uptake of beta carotene by the human gut and its conversion into vitamin A were quite inefficient, resulting in one vitamin molecule for every 12 molecules of beta carotene. Nobody knew the true rate at the time, but a recent, soon-to-be-published study among healthy volunteers who ate cooked golden rice, led by Robert Russell of Tufts University in Boston, suggests that it's more like one for every three or four. "That's really quite good," says Russell, who supports the golden rice project. (A similar study is planned among people with marginal vitamin A deficiency in Asia.)

Haerlin says his calculations were based on the best data at the time. But even if they were correct, Potrykus says, the first golden rice was just a proof of principle. Greenpeace might as well have blamed the Wright brothers for not building a transatlantic airplane, he says.

Industry Gets In

The low beta-carotene yield would eventually be tackled by Syngenta—even though Potrykus resented the way the company got involved. Between 1996 and 1999, Beyer's lab received funding through a European Commission contract that also included agrochemical giant Zeneca (called AstraZeneca after a merger in 1999). Under the program's rules, any benefits had to be shared by the signers. AstraZeneca had not worked on golden rice per se, Potrykus says, but the company claimed a share of that intellectual property anyway; it was interested in developing the technology commercially, for instance in health foods, says Potrykus, who was initially "furious" that a big corporation now had a say in his project.

David Lawrence has a different take on those events: At the time, AstraZeneca primarily wanted to support the humanitarian development of golden rice, says the current head of research at Syngenta; the company didn't have any commercial plans. (AstraZeneca's agribusiness division merged with that of Novartis to form Syngenta in 2000.) But whoever's right, the move proved a blessing in disguise, Potrykus now says. At Syngenta, he found a new partner in Adrian Dubock, a bubbly, fast-talking Brit with experience in patents, product development, regulation, and marketing—subjects Potrykus and Beyer admit they were clueless about.

Dubock helped work out a deal in which Syngenta could develop golden rice commercially, but farmers in developing countries who make less than $10,000 a year could get it for free. He also helped solve patent problems with several other companies. Dubock retired from Syngenta in 2007 but remains involved as a member of the Golden Rice Humanitarian Board, a group Potrykus chairs. "Without him, the project would have ended already," Potrykus says.

But perhaps most important, Syngenta scientists replaced a daffodil gene with a maize gene, thus creating a new version of golden rice, dubbed GR2, that produces up to 23 times more beta carotene in its seeds. Even with the one-in-12 conversion factor, that meant 72 grams of dry rice per day would suffice for a child, the company's scientists said in 2005. A 2006 paper by Alexander Stein of the University of Hohenheim in Stuttgart, Germany, estimated that the rice could have a major public health impact at a reasonable cost.

Those results didn't convince the skeptics. Real-world studies are still lacking, says WHO malnutrition expert Francesco Branca, noting that it's unclear how many people will plant, buy, and eat golden rice. He says giving out supplements, fortifying existing foods with vitamin A, and teaching people to grow carrots or certain leafy vegetables are, for now, more promising ways to fight the problem.

A Golden Future?

Today, the debate about golden rice has quieted down, in part because its inventors are keeping a low profile. Syngenta stopped its research on golden rice and licensed the rights to GR2 to the humanitarian board on World Food Day in 2004; given consumers' distrust, there was no money in it, says Lawrence. Most golden rice work is now taking place at six labs in the Philippines, India, and Vietnam, the countries chosen as the best candidates for the crop's launch.

There's a long way to go. Both the original golden rice, now called GR1, and GR2 were created with *Japonica* cultivars that are scientists' favorites but fare poorly in Asian fields. Researchers are now backcrossing seven GR1 and GR2 lines with the long-grained, nonsticky *Indica* varieties popular among Asia's farmers. In early April, researchers at the International Rice Research Institute in the Philippines finally started a field trial with a GR1 backcrossed into a widely used *Indica* variety called IR64—the first field trial ever in Asia. (The only other outdoor studies were two done in Louisiana in 2004 and 2005.) The new varieties must not only produce enough beta carotene but also pass muster in terms of yield, seed quality, and appearance.

The project could have been much further along, Potrykus says, if there weren't so many rules governing GM crops that make little sense. Conventional breeders can bombard plant cells with chemicals and radiation to create useful mutants without having to check how it affects their DNA; a GM insertion must be "clean"—that is, the extra genes must sit neatly in a row without disrupting other genes—which adds months or even years to the lab work. Because field trials take so long to get approved, researchers have been confined to greenhouses, in which they have trouble growing the large numbers required for breeding and feeding studies. These requirements have caused "year after year of delays," Potrykus complains.

Even if field trials are successful, there are no guarantees that golden rice will eventually be approved in the target countries. Use of other GM crops, such as Bt cotton, has exploded in Asia in recent years. But GM rice has languished. In India and China, regulatory agencies have shied away from approving insect-resistant GM rice despite extensive testing. "The expectation is that they will [be approved] eventually," says Toenniessen, "but it's a major decision for any Asian country." Thailand, a major rice exporter, has decided to steer clear of GM rice altogether.

Kavitha Kuruganti of the Centre for Sustainable Agriculture, an anti-GM group in Hyderabad, India, promises a major battle should golden rice head to the market in India. She thinks that the crop is unnecessary and probably unsafe to eat and that a massive switch would reduce diversity and threaten India's food security. "We will try to organize a broad public debate," she says.

Not Worth Funding?

Whether justified or not, the turmoil over golden rice has shaped other efforts to improve the nutritional value of crops. Take HarvestPlus. With a $14 million annual budget that targets 12 crops, it aims to boost levels of three key nutrients: vitamin A, iron, and zinc. It relies almost entirely on conventional breeding—which has Greenpeace's blessing—because it wants to have an impact fast, says Bouis, the director. What little GM technology HarvestPlus supports is a "hedge," in case the political and regulatory climates shift.

But in plants that have little or no natural ability to produce a nutrient, breeders have nothing to work with. Thus, vitamin A–enriched non-GM rice and sorghum are essentially off the table, says Bouis, as is boosting zinc and iron in sweet potatoes and cassava. Iron in rice is a question mark.

The uncertainty about the future of GM foods also tends to scare off the financial donors on which programs like HarvestPlus depend. Rockefeller, for instance, is frustrated that a GM rice whose field trials it helped pay for in China is stalled, says Toenniessen. "To avoid making the decision to approve it, the Chinese keep asking for more field trials," he says. "In the end, that becomes a foolish use of our funds."

The only charity still investing massively in GM crops with enhanced nutritional value is the Bill and Melinda Gates Foundation. Through its Grand Challenges in Global Health initiative, it is spending more than $36 million to support not only golden rice but also GM cassava, sorghum, and bananas. The foundation declined to comment for this story. But the researchers it supports say that they are optimistic that their products will make it through the pipeline.

James Dale of Queensland University of Technology in Brisbane, Australia, who heads a project to add iron, vitamin A, and vitamin E to bananas, says he has learned several lessons from golden rice, including the importance of local "ownership"—which is why he has teamed up with researchers in Kampala. "This will be a Ugandan banana made by Ugandans," he says.

Not that this mollifies opponents. Greenpeace will fight to keep GM bananas, cassava, and sorghum from poor countries' fields, just as it will keep opposing golden rice, says Janet Cotter of Greenpeace's Science Unit in London.

Battle-scarred, Potrykus says he hasn't given up hope that the regulatory system can be overhauled so that GM technology can benefit the poor. He believes a massive, multimillion-dollar information campaign might help convert the public. He has tried in vain to contact Bill Gates in hopes of tapping his wealth for such a media blitz.

He also wrote the late Pope John Paul II to ask for support for golden rice. "You know the definition of an optimist?" he jokes: "Someone who's asking the church for money." His Holiness declined, but Potrykus was invited to join the Pontifical Academy of Sciences, where he hopes to convene a meeting on golden rice next year—the 10th anniversary of his tarnished invention.

J. Madeleine Nash

Back to the Future

FROM *HIGH COUNTRY NEWS*

The Paleocene-Eocene Thermal Maximum took place 55 million years ago. But that carbon-dioxide-influenced rise in the Earth's mean temperature has implications for our own time, as researchers are discovering.

CLAMBERING ONTO A DUN-COLORED KNOLL, NOT FAR from the small town of Worland, Wyo., Scott Wing stares out at the deeply abraded hills that sweep towards him like the waves of a vast stony ocean. "That's it," he says, pointing to a sinuous ribbon of rose-colored rock. "That's the Big Red." I follow his gaze, noting how the Big Red snakes into an arroyo, then disappears around a bend. Even to my untrained eye, the geological band seems to glow with a fierce, otherworldly intensity.

In some places, Wing explains, the Big Red is composed of multiple stripes; in others, it wends through the landscape as a single line of color. Then, too, the capricious hand of erosion has exposed it here, left it hidden over there. But after hours of pondering the pieces of this jigsaw, Wing, a paleo-botanist at the Smithsonian's Museum of Natural History in Washington, D.C., believes he can now follow the Big Red for a distance of 25 miles, from the base of the solitary outcrop that looms in the distance all the way to the Sand Creek Divide.

The Sand Creek Divide is a high point in Wyoming's Big Horn Basin. From it you can see the emerald patchwork of irrigated sugar beet and malt barley fields that hug the Big Horn River as well as the jagged mountain ranges—the Absarokas, the Big Horns, the Owl Creeks—that define the edges of this harsh mid-latitude desert. Temperatures here regularly dip below 0 degrees Fahrenheit in wintertime and, in summer, soar well past 100. Away from waterways, the vegetation amounts to little more than a stippling of sagebrush intermingled with stands of invasive cheatgrass and ephemerally blooming wildflowers.

But between 55 and 56 million years ago, says Wing, the Big Horn Basin was a balmy, swampy Eden, teeming with flora and fauna that would be at home in today's coastal Carolinas. Crocodiles, turtles and alligator gar plied the waters of meandering rivers, and early mammals scampered through woodlands filled with the relatives of modern sycamores, bald cypresses and palms. And then, all of a sudden, things got a whole lot warmer. In a geological eye blink—less than 10,000 years, some think—global mean temperatures shot up by around 10 degrees Fahrenheit, jumpstarting a planetary heat wave that lasted for over 150,000 years.

Here, in the southeastern sector of the Basin, the Big Red is the most vivid marker of this exceptionally torrid time—the Paleocene-Eocene Thermal Maximum, or PETM, as most paleontologists call it. By following the Big Red, Wing and his colleagues hope to locate fossils and other clues that will help them reconstruct this long vanished world—a world with unexpected relevance for us as we hurtle towards our own rendezvous with climate change.

Scientists believe that, then as now, the earth warmed in response to a precipitous release of carbon dioxide and other heat-trapping gases, setting in motion events that reverberated through both marine and terrestrial ecosystems. But where did those gases come from so long ago? What triggered their sudden release? And, most important of all, how likely is it that the PETM, or something disquietingly close to it, could happen all over again?

In 1972, when a 17-year-old Wing made the first of many trips to the Big Horn Basin, scientists knew too little even to frame such questions. Today, however, dozens of paleontologists, oceanographers, geochemists and climate modelers are racing to come up with answers. Nowhere have they struck a more productive lode than in these candy-striped badlands. As Wing says, "You can literally walk up to a layer of rock and know that the Paleocene-Eocene boundary starts *here*."

Leaning on a long-handled shovel, Wing goes over the field schedule with a couple of colleagues, then heads back to Dino, a rust-colored 1970 Suburban with a bird-like dinosaur painted on each side. Wing bought this unlikely chariot in 1987 and somehow has kept it running ever since.

Five minutes later, he pulls up to the site that everyone refers to as "Ross's quarry" in honor of University of Nebraska paleontologist Ross Secord, who discovered it last year. Wing's crew has formed a conga line of shovelers, and as their 53-year-old leader scrambles up from below, they fling clouds of grit in his direction. Eventually, the pace of shoveling slows down so that promising chunks of rock can be individually examined and, if necessary, split open with a hammer. The best specimens are passed to Wing, who peers at each one through his eyepiece and decides whether to keep or discard it.

"This is a good one," Wing calls, so I climb up to see. On the surface of the rock is an exquisitely formed leaf, its veins and margins perfectly preserved. Grayish brown in color and slightly dank, the 55-million-year-old leaf looks like it might have fallen last week and is just now beginning to molder. Adding to the illusion of freshness, its fossilized tissue retains traces of the waxes that once comprised its protective exterior coating.

The plant to which this leaf once belonged, Wing thinks, migrated from far to the south in response to warming temperatures. Like a time capsule, the leaf carries information that can illuminate what it was like to live in a rapidly warming world.

"So far, what we've learned is that processes we're now affecting are so complicated that we can't easily model them," Wing says. "We can monitor them, but over short periods of time there's so much noise in the system that it overwhelms the signal. That's why the geological and paleontological record is so important. It's one of the few ways we can look into how the system works." With that, Wing turns away to squint at another leaf. Unshaven, with a broad-brimmed hat squashed onto his head and a notebook stuffed into a field vest pocket, he looks just like the seasoned fossil hunter he is.

EVEN BEFORE IT HAD A name, the Paleocene-Eocene Thermal Maximum was starting to fascinate Wing. For some time, it had been clear to paleontologists studying the evolution of mammals that the transition between the Paleocene and the Eocene was marked by the kind of innovative burst that implies sweeping ecological change. Yet no hint of such a change had appeared in any of the fossil leaves Wing had collected. He would stare at leaves from the Paleocene and leaves from the Eocene, but see almost no difference between them. "It was getting to be annoying," he recalls.

The Paleocene is the geological epoch that started 65 million years ago, right after a wayward asteroid or comet crashed into the planet, ending the reign of the dinosaurs. At the time, mammals were rather simple, general-purpose creatures with few specializations: Their teeth, their ankle bones and joints all look extremely primitive. Then, barely 10 million years later, at the dawn of the Eocene, the first relatives of deer abruptly appear, along with the first primates and first horses.

"You can literally draw a line through the rock," says Philip Gingerich, a vertebrate paleontologist at the University of Michigan. "Above it there are horses; below it there aren't." In fact, where Gingerich works—at Polecat Bench, in the northern sector of the Big Horn Basin—you can actually see the line, in the form of a band of light gray sandstone. Oddly enough, many fossil mammals commonly found above this line, including those first horses, were abnormally small. Typically, Gingerich says, Eocene horses grew to the size of modern-day cocker spaniels, but these horses were "about the size of Siamese cats."

In 1991, as Gingerich and others were marveling over the miniature mammals of Polecat Bench, oceanographers James Kennett of the University of California, Santa Barbara, and Lowell Stott of the University of Southern California investigated a major extinction of small, shelly creatures that, during the late Paleocene, lived on the sea floor off the coast of Antarctica. This massive die-off, they found, coincided with a steep rise in deep ocean temperatures and a curious spike in atmospheric carbon.

Less than a year later, paleontologist Paul Koch and paleo-oceanographer James Zachos, both now at the University of California, Santa Cruz, teamed up with Gingerich to show that this geochemical glitch had also left its calling card on land. The trio established this indirectly by measuring the carbon content of fossilized teeth and nodules plucked from the Big Horn Basin's 55.5-million-year-old rocks.

To Wing, it began to seem increasingly implausible that plant communities could have segued through the PETM unaffected. So in 1994, he started a methodical search for the fossils that he was all but sure he had missed, returning year after year to the Big Horn Basin. He started in its southeastern corner and then moved north to explore Polecat Bench and the Clarks Fork Basin. Yet it wasn't until 2003, when he reached the Worland area, that he began to meet with success.

At first, he found just a smattering of leaves, too few to suggest any pattern. Then, in 2005, at the end of a long day, he slid his shovel into a grayish mound and pulled out a tiny leaf. "I knew immediately that this was totally different from anything I'd seen before, that this was really dramatic, so I got down on my knees and poked the shovel in again and then again. In every shovelful, there were more leaves coming out. First I started to laugh; then I started to cry. And then I looked up."

Staring down at Wing was a new field assistant. "He had a look on his face that said, 'Now I'm going to die.' I understood what he must have been thinking. 'Here I am, I've just graduated from college. I've never camped before. I've never been on a paleontological expedition before. It's 6 o'clock in the evening. It's still 100 degrees. I don't know where I am. And it looks like the boss has gone completely nuts!' So I said to him, 'Really, it's OK. I'm not crazy. It's just that I've been looking for this since you were 10 years old!' "

From that one site, Wing went on to extract more than 2,000 leaf fossils representing 30 different species. Missing from the mix are the cypresses and other conifers that were so common during the Paleocene; gone also are the distant cousins of broadleaf temperate zone trees like sycamores, dogwoods, birches. In their place are the legumes, a family of plants, shrubs and trees—think of acacias and mimosas—that thrive today in seasonally dry tropical and subtropical areas.

"What you see is almost a complete changeover from what was growing here before," Wing marvels. "What this means is that you could have stood in this one spot in Wyoming, surrounded by a forest, and everything would have looked pretty much the same for millions of years. And then, over a few tens of thousands of years, almost all the plants you're familiar with disappear and are replaced by plants you've never seen before in your life." At least some of the newcomers migrated north from as far away as the Mississippi Embayment, precursor of the Gulf of Mexico. With them came a wave of small but voracious predators: Many of the fossil leaves are peppered with the scars left by chewing, sucking, mining and boring insects.

* * *

It's a cool, clear morning, with just a few wisps of cirrus streaking the sky, when Francesca Smith settles into a spot just above Ross's quarry, sitting cross-legged on the crackled ground. Across the road, along the ridgeline, are three petroleum pump jacks, their heads slowly rising and dipping as underground reservoirs empty, then fill with oil.

An associate professor at Northwestern University, the bubbly, brown-haired geochemist arrived only yesterday, having driven herself and two young assistants out from Illinois. Wing hands her a slab of mudstone. "Look," he says, pointing to the merest fragment of a leaf splayed across the surface. "That's a little piece of organic matter—it's probably part of the cuticle." At that moment, the wind picks up, and the paper-thin specimen peels back from the rock, threatening to fly away. "Emergency wrap!" Smith shouts, hastily enfolding her prize in an envelope of foam.

For a moment, Smith contemplates the pump jacks, visual metaphors linking the prehistoric world she and Wing are exploring to our world's present and future. The carbon released during the PETM, she notes, is also thought to have come from organic sources, just like the carbon we pump into the atmosphere every time we turn on a light or drive a car. "The only difference," Smith reflects, "is that we're doing it much, much faster."

During the Paleocene-Eocene Thermal Maximum, scientists estimate that a massive amount of carbon—4 to 5 trillion metric tons, perhaps—flooded into the atmosphere. That's about 10 times more carbon than humans have pumped out since 1751, and the rough equivalent of how much carbon remains stored in fossil fuels.

From a climatological perspective, it makes sense that the infusion of that much carbon would jack up temperatures. After all, carbon combines with oxygen to form carbon dioxide, which, next to water vapor, is the most abundant of the planet's greenhouse gases. As their name suggests, these gases (which also include methane and nitrous oxide) behave rather like the transparent panes of a greenhouse: They allow the sun's rays to stream in but trap a good deal of the heat the earth beams back in response.

In general, this is a good thing; it helps create what some call the Goldilocks effect—the fact that, to human beings and other creatures, Earth's temperature seems "just right," neither hellishly hot like Venus nor bitterly cold like Mars. That's not to say that our planet's pane of greenhouse gases never varies in thickness. Ancient air bubbles trapped in Antarctica's ice show that levels of carbon dioxide declined during past ice ages and rose during warm interglacials such as our own.

It's not clear how much carbon dioxide there was in the atmosphere on the eve of the PETM, but scientists think levels may have reached somewhere between 500 and 750 parts per million. This compares to 380 parts per million at present, 280 parts per million in pre-industrial times, and 180 during past glacial high stands. As a result, the late Paleocene was already quite warm, about as warm as many climatologists project our world could become by the start of the next century.

The large amount of carbon dioxide in the pre-PETM atmosphere almost certainly came from a sustained spate of volcanic eruptions. (During the Paleocene, volcanoes were particularly active.) As a result, the atmosphere's load of carbon dioxide gradually rose. Then, around 55.5 million years ago, carbon dioxide levels shot up very sharply, perhaps to 1,800 or more parts per million, leaving behind a distinctive geochemical signature.

The signature, Smith explains, takes the form of a dramatic shift in the ratio between two stable forms of carbon, heavier carbon-13 and lighter carbon-12. It's this shift that scientists first picked up in the calcareous shells of marine organisms, then found in the teeth of terrestrial mammals. Last year, Smith and Wing showed that, in the Big Horn Basin, the shift is captured by leaf waxes as well. "And there is only one way we know of to shift the ratio as much as it shifted," Smith says, "and that's to add a lot more light carbon."

The richest concentrations of light carbon are found in organic materials, including fossil fuels like coal (which forms from deeply buried plants) and methane gas (primarily a byproduct of microbial decomposition). While scientists are still not sure what triggered the massive release of light carbon at the start of the PETM, they do have a number of possible culprits, including fires that raged through forests, dried-up peat bogs and even underground coal seams, and effusively erupting volcanoes whose magma intruded into organic-rich sediments, cooking out the carbon.

Of all the scenarios so far floated, perhaps the most provocative invokes the dissolution of methane hydrates on the seafloor. These are ice-like solids in which water molecules form crystalline cages that entrap molecules of gas; they form, and remain stable, within specific ranges of temperature and pressure. At the end of the Paleocene, Rice University earth scientist Gerald Dickens has suggested, a jolt of warmth from an unknown source pushed these strange solids to the point that the methane gas inside them started burbling up through the ocean and into the atmosphere.

Methane is much shorter-lived than carbon dioxide, but it's also a more effective greenhouse gas. And as methane breaks down, the carbon it contains recombines with oxygen to form carbon dioxide, which can circulate through the climate system for thousands of years.

After serious study, many experts have concluded that not enough methane was locked up in hydrate form to have single-handedly caused the PETM. That does not constitute an absolution, however. A big release of seafloor methane could still have been part of a sustained chain reaction whereby an initial rise in carbon caused enough warming to trigger the release of additional carbon that caused still more warming, and so on. Might the carbon we are so heedlessly pumping out today spark a similar sequence of events?

As scientists try to imagine the consequences of our greenhouse gas emissions, they invariably return to this question. The earth abounds with "traps" for carbon—not just seafloor hydrates, but also terrestrial forests, marine plankton and frozen Arctic soils. Some of these our own warming climate may already be springing open. Scientists from the University of Alaska, Fairbanks, recently calculated that the permafrost of the far North sequesters 100 billion tons of carbon in its top three feet alone; as the permafrost thaws, that carbon will progressively leak into the atmosphere.

The release of carbon from some of these traps will be offset by the uptake of carbon in others. But at some point, scientists worry, the release of carbon may so far outweigh its absorption that the situation will cascade out of control. By the time we realize we're in serious trouble, in other words, it may be too late to do much about it.

INSIDE A CAVERNOUS TENT BATHED in golden late-afternoon light, the resident team of vertebrate paleontologists pores over the day's haul, emitting sporadic whoops of surprise. "That might be a eureka," exclaims Yale University graduate student Stephen Chester, peering through a microscope at a tooth embedded in a jawbone fragment. "That might be a primate."

"It is! It's totally *Teilhardina*!" Doug Boyer, a Ph.D. candidate at New York's Stony Brook University, enthusiastically agrees. *Teilhardina*, he explains, is the Latin name for a group of primates that appear in Asia, Europe and North America at roughly the same time.

"That's a horse, Douggie," Chester says, examining another tooth. It belongs to *Hyracotherium sandrae*, the unusually small species first identified at Polecat Bench. The tooth is a shiny dark amber and it's very tiny. Why was this horse, *Hyracotherium sandrae*, so small?

The most straightforward explanation is that *Hyracotherium sandrae* was simply a small-sized species that migrated into the Big Horn Basin from somewhere else. But the University of Michigan's Gingerich champions a more intriguing possibility. He suggests that its diminutive stature could be the consequence of a decline in available nutrients, notably protein. Horticultural experiments have shown that some plants, when bathed in high concentrations of carbon dioxide, have less protein in their leaves.

The same phenomenon may also have caused insects to become more voracious, which is consistent with the leaf damage displayed by Wing's fossils. Here again, however, there are other possible explanations. For example, higher temperatures alone would have raised insect food requirements by quickening metabolic rates and encouraging year-round breeding. "During the PETM, many things are changing all at once, and it's hard to separate one from another," Wing observes.

At present, Wing is in the field camp's cooking tent, heating up his favorite utensil, a big black wok that can handle dinner for 16. Just behind him, seated at a long metal table, Mary Kraus, a wiry sedimentologist from the University of Colorado, is starting to peel a big pile of russet potatoes. She and her daughter, Christina, have had a great day, she says, digging a trench through the pastel paleosols of the badlands surrounding a perennially dry fork of Nowater Creek. "Look at this treasure," she exclaims, holding up a fossilized insect burrow shaped vaguely like a cowboy's boot.

Like tree rings and deep sea sediments, burrows are what scientists refer to as "proxies." Simply put, proxies are natural systems that record and preserve information about past climates, not unlike modern instruments. Crayfish burrows indicate soils that experience large fluctuations in wetness; earthworm and beetle burrows suggest drier conditions. The colors of ancient soils are also proxies. For example, the degree of redness—whether the color tends towards orange or towards purple—can be correlated with specific ranges of soil moisture.

There are many other types of proxies, including fossil leaves, teeth and the shells of marine organisms. Typically these proxies record shifts in the ratio between heavy and light elements. Oxygen shifts can be translated into temperature; hydrogen shifts into relative humidity. Changes in leaf size and shape can likewise be read as proxies for temperature and moisture, though as Wing admits, "We don't fully understand why."

Multiple proxies, Wing says, suggest that, during the PETM, this area of Wyoming was rather similar to South Florida, with a mean annual temperature of around 75 degrees F and annual precipitation between 30 and 55 inches. The precipitation may have followed a strongly seasonal pattern, especially towards the beginning, with part of the year being quite dry, Wing believes. But that's just the broad-brush picture. Wing, Smith and Kraus, along with University of Florida paleontologist Jonathan Bloch, who heads up the vertebrate fossil collection effort, are working on reconstructing the regional climate in much finer detail.

Kraus is using ancient soils to begin mapping what seems like a climatological progression. The initial phase of the PETM looks rather dry, she says, and the middle phase appears drier still, though there are signs of very rapid soil deposition from flooding along rivers and streams. Towards the end of the PETM, in the Big Red itself, she is finding hints that conditions may have become wetter. Among other things, the rocks of the Big Red contain a lot of purple, a color suggestive of higher water tables and more poorly drained soils.

The Big Horn Basin, Kraus says, is probably the ideal place to try to pull together a comprehensive picture of how climate changed on a regional scale over the course of the PETM. "Where else can you go and find 5,000-year intervals stacked one on top of another?" she asks. "Where else can you go and know that 40 meters of rock (about 130 feet) equals 150,000 years?" That's around how long it took the PETM to wind up and wind down, so it's not surprising that, over the course of so many millennia, both regional and global climate patterns underwent successive changes. The wind-up, of course, was the fast part; it was the wind-down that took a long time.

DELICIOUS AROMAS ARISE FROM THE wok as Wing adds onions and garlic and ginger. Following their noses, the vertebrate fossil crew streams in. Soon we are all sitting in camp chairs, chowing down on rice and spicy curry. After dinner, when the dishes are all washed, dried and put away, Wing pulls out the battered acoustic guitar he bought for $8 years ago in a Worland pawn shop. He starts strumming it softly. Two more members of the group join in, one on a battery-powered keyboard, the other on a tinny guitar.

As the trio warbles out a medley of familiar songs, I contemplate the gossamer sash of the Milky Way as it flows across the nighttime sky. The universe is almost 14 billion years old. Some of the stars in our galaxy are 10 billion years old. The earth and the sun it circles are around 4.5 billion years old. Measured against such a long stretch of time, the duration of the Paleocene-Eocene Thermal Maximum seems absurdly insignificant. Compared to our own allotment of some few score years and ten, however, it looms a great deal larger. In little more than 150,000 years, ice ages came and went and started anew. In 150,000 years, modern humans diverged from their archaic ancestors and began to spread across the world.

For most of that time, our forebears lived in small, mobile clusters of hunter-gatherers. They began coalescing into settled agricultural communities perhaps 10,000 years ago. Their history as members of a technologically advanced industrial civilization is breathtakingly recent, powered into existence by the 18th-century invention of efficient coal-fired steam engines. Who would have predicted that over the span of so few centuries, the clever, adaptable descendants of Eocene primates would become so numerous—and so dependent on carbon-based fuels—as to unbalance the planet?

Perhaps, many hundreds of thousands of years from now, paleontologists from some advanced civilization will uncover fossils from our world and marvel at the carbon shift recorded by the teeth of free-ranging cattle and sheep, and the leaves of garden shrubs and trees. Will those beings fathom the real wonder of our story, the fact that we had glimmers of what the future held and yet failed to use that knowledge? Or will our story, like one of Shakespeare's dark comedies, work its way to a happier ending?

It's not that the PETM offers a precise road map to our future, Wing says, when I ask for his thoughts. "It's more that it's an example of the surprises that are waiting for us out there. How was it that Mark Twain put it? History does not repeat itself, but it sure does rhyme."

ELIZABETH ROYTE

A Tall, Cool Drink of . . . Sewage?

FROM *THE NEW YORK TIMES MAGAZINE*

With water shortages looming, entrepreneurs and researchers alike are looking for ways of preserving and extending the Earth's supply of fresh water. Elizabeth Royte travels to California to give "recycled water" a try.

BEFORE I LEFT NEW YORK FOR CALIFORNIA, WHERE I planned to visit a water-recycling plant, I mopped my kitchen floor. Afterward, I emptied the bucket of dirty water into the toilet and watched as the foamy mess swirled away. This was one of life's more mundane moments, to be sure. But with water infrastructure on my mind, I took an extra moment to contemplate my water's journey through city pipes to the wastewater-treatment plant, which separates solids and dumps the disinfected liquids into the ocean.

A day after mopping, I gazed balefully at my hotel toilet in Santa Ana, Calif., and contemplated an entirely new cycle. When you flush in Santa Ana, the waste makes its way to the sewage-treatment plant nearby in Fountain Valley, then sluices not to the ocean but to a plant that superfilters the liquid until it is cleaner than rainwater. The "new" water is then pumped 13 miles north and discharged into a small lake, where it percolates into the earth. Local utilities pump water from this aquifer and deliver it to the sinks and showers of 2.3 million customers. It is now drinking water. If you like the idea, you call it indirect potable reuse. If the idea revolts you, you call it toilet to tap.

Opened in January, the Orange County Groundwater Replenishment System is the largest of its type in the world. It cost $480 million to build, will cost $29 million a year to run and took more than a decade to get off the ground. The stumbling block was psychological, not architectural. An aversion to feces is nearly universal, and as critics of the process are keen to point out, getting sewage out of drinking water was one of the most important public health advances of the last 150 years.

Still, Orange County forged ahead. It didn't appear to have a choice. Saltwater from the Pacific Ocean was entering the county's water supply, drawn in by overpumping from the groundwater basin, says Ron Wildermuth, who at the time we talked was the water district's spokesman. Moreover, population growth meant more wastewater, which meant building a second sewage pipe, five miles into the Pacific—a $200 million proposition. Recycling the effluent solved the disposal problem and the saltwater problem in one fell swoop. A portion of the plant's filtered output is now injected into the ground near the coast, to act as a pressurized barrier against saltwater from the ocean. Factor in Southern California's near chronic drought, the county's projected growth (another 300,000 to 500,000 thirsty people by 2020) and the rising cost of importing water from the Colorado River and from Northern California (the county pays $530 per acre-foot of imported water, versus $520 per acre-foot of reclaimed water), and rebranding sewage as a valuable resource became a no-brainer.

With the demand for water growing, some aquifers dropping faster than they're replenished, snowpacks thinning and climate change predicted to make dry places even drier, water managers around the country, and the world, are contemplating similar schemes. Los Angeles and San Diego, which both rejected potable reuse, have raised the idea once again, as have, for the first time, DeKalb County, Ga., and Miami-Dade County, Fla.

While Orange County planned and secured permits, public-relations experts went into overdrive, distributing slick educational brochures and videos and giving pizza parties. "If there was a group, we talked to them," says Wildermuth, who recently left Orange County to help sell Los Angelenos on drinking purified waste. "Historical societies, chambers of commerce, flower committees." The central message was health and safety, but the persuaders didn't skimp on buzz phrases like "local control" and "independence from imported water." Last winter, the valve between the sewage plant and the drinking-water plant whooshed open, and a new era in California's water history began.

WHEN I VISITED THE PLANT, a sprawl of modern buildings behind a concrete wall, in March, Wildermuth, in a blue sport coat and bright tie, acted as my guide. "Quick!" he shouted at one point, mounting a ledge and clinging to the rail over a microfiltration bay. "Over here!" I clambered up just as its contents finished draining from the scum-crusted tank. The sudsy water, direct from the sewage-treatment plant, was the color of Guinness. "This is the most exciting thing you'll see here, and I didn't want you to miss it," he said.

Wildermuth went on to explain what we were looking at: inside each of 16 concrete bays hangs a rack of vertical tubes stuffed with 15,000 polypropylene fibers the thickness of dental floss. The fibers are stippled with holes 1/300th the size of a human hair. Pumps pull water into the fibers, leaving behind anything larger than 0.2 microns, stuff like bacteria, protozoa and the dread "suspended solids."

The excitement and the bubbles were backwash: every 21 minutes, air is injected into the microfibers to blast them clean. The schmutz goes back to the sewage-treatment plant, and the cleaner water, now the color of chamomile tea, is pumped toward reverse-osmosis filters in another building. Before we saw that process, Wildermuth led me underground to inspect several enormous pumps and pipes large enough to crawl through. I noted that everything was clearly labeled and scrupulously clean. Then it dawned on me: reassurance was the reason we'd taken the detour.

We followed the pipes up to a sunlit, metal-clad building where the water, now dosed with an antiscalant and sulfuric acid to lower its pH, was forced at high pressure through hundreds of white tubes filled with tightly spiraled sheets of plastic membranes. Reverse osmosis, Wildermuth says, stops cold almost all nonwater molecules (things like salts, viruses and pharmaceuticals). The stuff that's removed is washed back to a pipe that discharges into the ocean. The filtered water, now known as permeate, moves one building over, where it's spiked with hydrogen peroxide, a disinfectant, and then circulated past 144 lamps emitting ultraviolet light. "Destruction of compounds through photolysis," Wildermuth said, nodding. Anything that's alive in this water can no longer reproduce.

Strolling back through the campus, Wildermuth took me to a three-part demonstration sink with faucets streaming. The basin on the right contained reverse-osmosis backwash: it was molasses black, topped with a rainbow slick of oil. "Don't touch," Wildermuth warned as I leaned in for a better look at the ocean-bound rejectamenta. The middle basin contained the chamomile water from microfiltration. And on the left was the stuff Orange County would eventually drink. It was clear and had no smell.

But even this suctioned, sieved and irradiated water wasn't quite set for sipping; it still needed to be decarbonized and dosed with lime, to raise its pH. Finally it would enter a massive purple pipe, which dives into the ground inside a nearby pump house and reappears 13 miles to the north, in Anaheim. There, the water would pour into Kraemer Basin, a man-made reservoir, where it would mix with the lake water and filter for six months through layers of sand and gravel hundreds of feet deep before utilities throughout the county pumped it into taps.

The reservoir is a prosaic ending for a substance that's been through the glitziest of technological wringers, transformed from sewage to drinking water only to be humbly redeposited into the earth. This final filtering step isn't necessary, strictly speaking, but our psyches seem to demand it.

To UNDERSTAND THE BASICS OF contemporary water infrastructure is to acknowledge that most American tap water has had some contact with treated sewage. Our wastewater-treatment plants discharge into streams that feed rivers from which other cities suck water for drinking. By the time New Orleans residents drink the Mississippi, the water has been in and out of more than a dozen cities; more than 200 communities, including Las Vegas, discharge treated wastewater into the Colorado River. That's the good news. After heavy rains, many cities discharge untreated sewage directly into waterways—more than 860 billion gallons of it a year, according to the Environmental Protection Agency. However—and this is where we can take solace—the sewage is massively diluted, time and sunlight help to break down its components and drinking-water plants filter and disinfect the water before it reaches our taps. The E.P.A. requires utilities to monitor pathogens, and there hasn't been a major waterborne-disease outbreak in this country since 1993. (Though there have been 85 smaller outbreaks between 2001 and 2006.)

So confident are engineers of so-called advanced treatment technologies that several communities have been discharging highly treated wastewater directly into reservoirs for years. Singapore mixes 1 percent treated wastewater with 99 percent fresh water in its reservoirs. (In Orange County, the final product will contain 17 percent recycled water.) Residents of Windhoek, Namibia, one of the driest places on earth, drink 100 percent treated wastewater. For 30 years, the Upper Occoquan Sewage Authority, in Virginia, has been mixing recycled wastewater with fresh water in a reservoir and serving it to more than a million people. Still, no system produces as much recycled water as Orange County (currently 70 million gallons a day, going up to 85 million by 2011), and none inserts as many physical and chemical barriers between toilet and tap.

Environmentalists, river advocates and California surfers—the sort of people who harbor few illusions about the purity of our rivers and oceans—generally favor water recycling. It beats importing water on both economic and environmental grounds (about a fifth of California's energy is used to move water from north to south). "The days are over when we can consider wastewater a liability," says Peter Gleick, president of the Pacific Institute, an environmental research group in Oakland. "It's an asset. And that means figuring out how best to use it."

As we deplete the earth's nonrenewable resources, like oil and metals, the one-way trip from raw material to disposed and forgotten waste makes less and less sense. Already we recycle aluminum to avoid mining, compost organic material to avoid generating methane in landfills and turn plastic into lumber. As it becomes more valuable, water will be no different.

"We have to treat all waste as a resource," Conner Everts, executive director of the Southern California Watershed Alliance, says. "Our water source, hundreds of miles away, is drying up. If the population is growing, what are our options?"

Water conservation could take us a long way, as would lower water subsidies for farmers. But sooner or later, stressed-out utility managers come back to the same idea: returning wastewater to the tap.

The process isn't risk-free. Some scientists are concerned that dangerous compounds or undetectable viruses will escape the multiple physical and chemical filters at the plant. And others suggest that the potential for human error or mechanical failure—clogged filters or torn membranes that let pathogens through, for example—is too great to risk something as basic to public health as drinking water.

Recycled water should be used only as nondrinking water, says Philip Singer, the Daniel Okun Distinguished Professor of Environmental Engineering at the University of North Carolina. "It may contain trace amounts of contaminants. Reverse osmosis and UV disinfection are very good, but there are still uncertainties."

And then there are those whose first, and final, reaction is "yuck."

"Why the hell do we have to drink our own sewage?" asks Muriel Watson, a retired schoolteacher who sat on a California water-reuse task force and founded the Revolting Grandmas to fight potable reuse. She toured the Orange County plant but came away unsatisfied. "It's not the sun and the sky and a roaring river crashing into rocks"—nature's way of purifying water. "It's just equipment."

THE SANTA ANA RIVER FORMS in the San Bernardino Mountains and flows southwest through Riverside and then Orange counties to the sea, the largest coastal stream in Southern California. But that's not saying much: in the summer, the Santa Ana's flow is nearly 100 percent wastewater. The river's base flow—what enters the channel from runoff, rain and wastewater-treatment plants—is increasing. Not only is more effluent entering the river, a consequence of population growth, but as the county develops and paves more surfaces, rainwater runs off the earth faster, sluicing into the river channel before it can sink into the earth and replenish aquifers.

To capture and clean that water, the Orange County Water District has gone into hyper-beaver mode on the river. Twenty miles upstream from Anaheim, the water district has created the Prado Wetlands. It's a lovely place, lush with willow and mule fat, busy with butterflies and, over the course of the year, 250 species of birds. Moving through a series of rectangular ponds, river water filters slowly through thickets of cattails and bulrushes meant to extract excess nitrate from upstream dairy farms and sewage-treatment plants. Returned to the main channel, the water wends around T- and L-shaped berms that slow the water and maximize its contact with the river bottom. Gates and sluiceways then shunt the water into nine man-made ponds and pits. The goal is to get more water into the county's groundwater basin, a 350-square-mile, 1,500-foot-deep bathtub of sand and gravel layers, which act as natural scrubbers. The system upriver—using gravity and gravel—and the system in Fountain Valley—in tanks and tubes—both achieve the same goal. Sort of.

It's one of the many paradoxes of indirect potable reuse that the water leaving the plant in Fountain Valley is far cleaner than the water that it mingles with. Yes, the water entering the sewage-treatment plant in Fountain Valley is 100 percent wastewater and has a T.D.S.—a measure of water purity, T.D.S. stands for total dissolved solids and refers to the amount of trace elements in the water—of 1,000 parts per million. But after microfiltration and reverse osmosis, the T.D.S. is down to 30. (Poland Spring water has a T.D.S. of between 35 and 46.) By contrast, the "raw" water in the Anaheim basins has a T.D.S. of 600.

If everything in the Fountain Valley plant is in perfect working order, its finished water will contain no detectable levels of bacteria, pharmaceuticals or agricultural and industrial chemicals. The same can be said of very few water sources in this country. But once the Fountain Valley water mingles with the county's other sources, its purity goes downhill. Filtering it through sand and gravel removes some contaminants, but it also adds bacteria (not necessarily harmful, and local utilities will eventually knock them out with chlorine) and possibly pharmaceuticals.

In other words, nature messes up the expensively reclaimed water. So why stick it back into the ground? "We do it for psychological reasons," says Adam Hutchinson, director of recharge operations for the water district. "In the future, people will laugh at us for putting it back in, instead of just drinking it."

Psychologists and marketers have spent a lot of time trying to figure out what makes a product, or a process, seem natural. Obviously, framing the issue properly is the key to acceptance. "If people connect the history of their water to contamination, you'll get a disgust response no matter how you treat that water in between," says Brent Haddad, an associate professor of environmental studies at the University of California at Santa Cruz. "But if you enable people to frame out that history by telling them, for example, that 'the clean water has been separated from the polluted water,' they no longer make that connection." We abridge history all the time, Haddad adds. "Think of the restaurant fork that was in the mouth of someone with a contagious disease, the pillow that was underneath people doing private adult things in a hotel bedroom. If you think of it that way, the intermediate steps, like washing with hot water, don't matter."

ALL WATER ON EARTH IS recycled: the same drops that misted Devonian ferns and dripped from the fur of woolly mammoths are watering us today. From evaporation to condensation and precipitation, the cycle goes on and on. But in the planet's drier regions, where the population continues to rise, we can expect the time between use and reuse to grow ever shorter, with purification, pipes and pumps standing in for natural processes. Instead of sand and gravel filtering our drinking water, microfibers and membranes will do the job; instead of sunlight knocking out parasites, we'll plug in the UV lamps.

You could argue that in coming to terms with wastewater as a resource, we'll take better care of our water. At long last, the "everything is connected" message, the bedrock of the environmental movement, will hit home. In this view, once a community is forced to process and drink its toilet water, those who must drink it will rise up and change their ways. Floor moppers will switch to biodegradable cleaning products. Industry will use nontoxic material. Factory farms will cut their use of antibiotics. Maybe we'll even stop building homes in the desert.

But these situations are not very likely. No one wants to think too hard about where our water comes from. It's more likely that the virtuosity of water technology will let polluters off the hook: why bother to reduce noxious discharges if the treatment plant can remove just about anything? The technology, far from making us aware of the consequences of our behavior, may give us license to continue doing what we've always done.

The recycled water coming out of the sink at the Fountain Valley plant looked good enough to drink. Wildermuth didn't press me to taste it, but I was eager for a sample—to satisfy my curiosity, and to be polite. I filled a plastic cup and took a sip. The water tasted fine, if a little dry; I'm used to something with more minerals. It did cross my mind that any potential health issues from drinking so-far undetectable levels of contaminants would be cumulative and take decades to manifest.

Then I reminded myself: no naturally occurring water on earth is absolutely pure. And most everything that's in Orange County's reclaimed water is in most cities' drinking water anyway.

It was hot, my throat was parched, and I asked for a refill.

DAVID QUAMMEN

Contagious Cancer

FROM *HARPER'S*

Cancer is already terrifying enough. But David Quammen discovers that some forms of it are contagious—and give scientists a chance to watch evolution in action.

DURING THE EARLY MONTHS OF 1996, NOT LONG before Easter, an amateur wildlife photographer named Christo Baars made his way to the Australian island-state of Tasmania, where he set up camp in an old airport shack within the boundaries of Mount William National Park. Baars's purpose, as on previous visits, was to photograph Tasmanian devils, piglet-size marsupials unique to the island's temperate forests and moors. Because devils are nocturnal, Baars equipped his blind with a cot, a couple of car batteries, and several strong spotlights. For bait he used road-kill kangaroos. Then he settled in to wait.

The devil, known to science as *Sarcophilus harrisii*, lives mostly by scavenging and sometimes by predation. It will eat, in addition to kangaroo meat, chickens, fish, frogs, kelp maggots, lambs, rats, snakes, wallabies, and the occasional rubber boot. It can consume nearly half its own body weight in under an hour, and yet—with its black fur and its trundling gait—it looks like an underfed bear cub. Fossil evidence shows that devils inhabited all of Australia until about 500 years ago, when competition with dingoes and other factors caused them to die out everywhere but in Tasmania, which dingoes had yet to colonize. More recently, Tasmanian stockmen and farmers have persecuted devils with the same ferocity directed elsewhere at wolves and coyotes. The devils' reproductive rate, opportunistic habits, and tolerance for human proximity, however, have allowed localized populations to persist or recover, and at the time of Baars's 1996 visit, their total number was probably around 150,000.

On his earlier visits, Baars had seen at least ten devils every night, and they were quick to adjust to his presence. They would walk into his blind, into his tent, into his kitchen, and he could recognize returning individuals by the distinctively shaped white patches on their chests. This trip was different. On the first night, his bait failed to attract a single devil, and the second night was only a little better. He thought at first that maybe the stockmen and farmers had finally succeeded in wiping them out. Then he spotted a devil with a weird facial lump. It was an ugly mass, rounded and bulging, like a huge boil, or a tumor. Baars took photographs. More devils wandered in, at least one of them with a similar growth, and Baars took more pictures. This was no longer wildlife photography of the picturesque sort; it was, or anyway soon would become, forensic documentation.

Back in Hobart, Tasmania's capital, Baars showed his pictures to Nick Mooney, a veteran officer of Tasmania's Parks and Wildlife Service who has dealt with the devil and its enemies for decades. Mooney had never seen anything like this. The lumps looked tumorous, yes—but what sort of tumor? Mooney consulted a pathologist, who suggested that the devils might be afflicted with lymphosarcoma, a kind of lymphatic cancer, maybe caused by a virus passed to the devils from feral cats. Such a virus might also be passed from devil to devil, triggering cancer in each.

More evidence of contagion began to accumulate. Three years after Baars shot his photographs, a biologist named Menna Jones took note of a single tumor-bearing animal, something she had not seen before. Then, in 2001, at her study site along Tasmania's eastern coast, her traps yielded three more devils with ulcerated tumors. That really got her attention. She euthanized the animals and brought them to a lab, where they became the first victims to be autopsied by a veterinary pathologist. The "tumors" (until then the term had been only a guess or a metaphor) did seem to be cancerous malignancies, but not of the sort expected from a lymphosarcoma-triggering virus. This peculiarity raised more questions than it answered. Tasmanian devils in captivity were known to be quite susceptible to cancer, at least in some circumstances, possibly involving exposure to carcinogens. But the idea that the cancer itself was contagious seemed beyond the realm of possibility. And yet, during the following year, Menna Jones charted the spread of the problem across northern Tasmania. Nick Mooney, meanwhile, had done some further trapping himself. At a site in the northern midlands, he captured twenty-three devils, seven of which had horrible tumors. Shocked and puzzled, he remembered the Baars photos from years earlier.

Further trapping (more than a hundred animals, of which 15 percent were infected) showed Mooney what Jones had also seen: that the tumors were consistently localized on faces, filling eye sockets, distending cheeks, making it difficult for the animals to see or to eat. Why faces? Maybe because devils suffer many facial and mouth injuries—from chewing on brittle bones, from fighting with one another over food and breeding rights, from the rough interactions between male and female when they mate. The bigger tumors were crumbly, like feta cheese. Could it be that tumor cells, broken off one animal, fell into the wounds of another, took hold there, and grew? This prospect seemed outlandish, but the evidence was leading inexorably to a strange and frightening new hypothesis: the cancer itself had somehow become contagious.

Under ordinary circumstances, cancer is an individuated phenomenon. Its onset is determined partly by genetics, partly by environment, partly by entropy, partly by the remorseless tick-tock of time, and (almost) never by the transmission of some tumorous essence. It arises from within (usually) rather than being imposed from without. It pinpoints single victims (usually) rather than spreading through populations. Cancer might be triggered by a carcinogenic chemical, but it isn't itself poisoning. It might be triggered by a virus, but it isn't fundamentally viral. Cancer differs also from heart disease and cirrhosis and the other lethal forms of physiological breakdown; uncontrolled cell reproduction, not organ dilapidation, is the problem.

Such uncontrolled reproduction begins when a single cell accumulates enough mutations to activate certain growth-promoting genes (scientists call them oncogenes) and to inactivate certain protections (tumor suppressor genes) that are built into the genetic program of every animal and plant. The cell ignores instructions to limit its self-replication, and soon it becomes many cells, all of them similarly demented, all bent on self-replication, all heedless of duty and proportion and the larger weal of the organism. That first cell is (almost always) a cell of the victim's own body. So cancer is reinvented from scratch on a case-by-case basis, and this individuation, this personalization, may be one of the reasons that it seems so frightening and solitary. But what makes it even more solitary for its victims is the idea, secretly comforting to others, that cancer is never contagious. That idea is axiomatic, at least in the popular consciousness. *Cancer is not an infectious disease.* And the axiom is (usually) correct. But there are exceptions. Those exceptions point toward a broader reality that scientists have begun to explore: Cancers, like species, evolve. And one way they can evolve is toward the capacity to be transmitted between individuals.

Devil tumor isn't the only form of cancer ever to achieve such a feat. Other cases have occurred and are still occurring. The most notable is Canine Transmissible Venereal Tumor (CTVT), also called Sticker's sarcoma, a sexually transmitted malignancy in dogs. Again, this is not merely an infectious virus that tends to induce cancer. The tumor cells themselves are transmitted during sexual contact. CTVT is widespread (though not common) and has been claiming dogs around the world at least since a Russian veterinarian named M. A. Novinsky first noted it in 1876. The distinctively altered chromosome patterns shared by the cells of CTVT show the cancer's lineal continuity, its identity across space and through time. Tumor cells in Dog B, Dog C, Dog D, and Dog Z are more closely related to one another than those cells are to the dogs they respectively inhabit. In other words, CTVT can be conceptualized as a single creature, a parasite (and not a *species* of parasite, but an *individual*), which has managed to spread itself out among millions of different dogs. Research by molecular geneticists suggests the tumor originated in a wolf, or maybe an East Asian dog, somewhere between 200 and 2,500 years ago, which means that CTVT is probably the oldest continuous lineage of mammal cells presently living on Earth. The dogs may be young, but the tumor is ancient.

Unlike devil tumor—now known as Devil Facial Tumor Disease, or DFTD—CTVT is generally not fatal. It can be cured with veterinary surgery or chemotherapy. In many cases, even without treatment, the dog's immune system eventually recognizes the CTVT as alien, attacks it, and clears it away, just as our own immune systems eventually rid us of warts.

The case of the Syrian hamster is more complicated. This tumor arose around 1960, when researchers at the National Cancer Institute, in Bethesda, Maryland, performed an experiment in which they harvested a naturally occurring sarcoma from one hamster and injected those cells (as cancer scientists often do) into healthy animals. When the injected hamsters developed malignancies, more cells were harvested. Each such inoculation-and-harvest cycle is called a passage. The experiment involved a dozen such passages, and over time the tumor began to change. It had evolved. The later generations, unlike the first, represented a sort of super tumor, capable of getting from hamster to hamster without benefit of a needle. The researchers caged ten healthy hamsters together with ten cancerous hamsters and found that nine of the healthy animals acquired tumors through social contact. The hamster tumor had leapt between animals—or anyway, it had been smeared, spat, bitten, and dribbled between them. (The tenth hamster got cannibalized before it could sicken.) In a related experiment, the tumor even passed between two hamsters separated by a wire screen. The scientists had in effect created a laboratory precursor of what would eventually afflict Tasmanian devils in the wild: a Frankenstein malignancy, a leaping tumor, which could conceivably kill off not just individuals but an entire species.

EARLY LAST SUMMER I WENT to Tasmania, where I met Menna Jones for an excursion to the Forestier Peninsula, a long hook of land that juts southeastward into the Tasman Sea. Jones supervises an experimental trapping program aimed at ridding the peninsula of tumor disease or, at least, determining whether that goal is achievable. The Forestier is a good place for such trials because the peninsula (and its lower extension, a second lobe called the Tasman Peninsula) is connected to the rest of Tasmania by only a narrow neck—just a two-lane bridge across a canal. If the disease could be eradicated from the entire peninsula, by removing all sick animals and leaving the healthy ones, Forestier and Tasman might be protected from re-infection by a devil-proof barrier across the bridge; and if that worked, the protected population could rebound quickly. The Forestier Peninsula, full of good habitat, might become a vital refuge for the species. Those measures might even validate a method—defense by tourniquet—that could be used on some of Tasmania's other peninsular arms.

Jones, who is a brisk, cordial woman with a mane of brown hair, picked me up in an official state Land Cruiser, and as she drove she described the effects seen so far. Her field people had culled more than a hundred devils within the past four months, she said, and though the size of the Forestier population seemed to be holding steady, the demographics had changed. Mature adults, the four- and five-year-olds, were being lost, and so three-year-olds, adolescents, were accounting for most of the parenthood. The biting associated with breeding brings fatal disease, and the disease kills fast—sex equals death, a bad equation for any species. "We think extinction is a possibility within twenty-five years," Jones said.

We crossed the little bridge onto the peninsula and, after a short drive through rolling hills of eucalyptus forest, rendezvoused with the trapping crew. The chief trapper was a young woman named Chrissy Pukk, Estonian by descent, Aussie by manner, wearing a pair of blue coveralls, a dangling surgical mask, and a leather bush hat. She had been trapping devils here for three years. Jones and I tagged along as Pukk and two volunteers worked a line of forty traps placed throughout the forest. The catch rate was high, and most of the captured devils had been caught previously and injected with small electronic inserts for identification. These devils came in on a regular basis, as if the traps were soup kitchens, and Pukk recognized many of them on sight. She and only she handled the animals, cooing to them calmingly while she took their measurements, checked their body condition, and, most crucially, examined their faces for injuries and signs of tumor. One devil, a robust male Pukk called Captain Bligh, showed wounds from a recent mating session: broken teeth, a torn nose, a half-healed cut below his jaw, and a suppurating pink hole on the top of his snout, deep enough that it might have been made by a melon scoop. But he seemed to be clean of tumor. "He's just a brawler," Pukk said. She released him, and he skittered off into the brush.

"You see a lot of old friends come and go," Pukk told me as she examined another animal. For instance, there was one she had caught the day before, a male called Noddy. She had last trapped him less than a week earlier, noted inflamed whisker roots, and released him; but in the brief passage of days, those inflammations had become tumors, and now Noddy was awaiting his fate in a holding trap.

Colette Harmsen, a veterinarian who had made the long drive south from Tasmania's Animal Health Laboratory at Mount Pleasant, was there to euthanize and autopsy any animal Pukk found unfit for release. She wore her black hair cut short, her jeans torn at the knee, a lacey black dress over the jeans, a black T-shirt reading SAVE TASSIE'S FORESTS over the dress, and, over it all, a pale blue disposable surgical smock. She was waiting at her pickup truck along with her pit bull, Lily, and her pet rat, CC, when Pukk arrived to deliver the unfortunate Noddy. Pukk and her crew returned to the trap line, Menna Jones went back to Hobart, and I stayed to watch Harmsen work. I had never seen anyone cut open a Tasmanian devil.

Her working slab was the tailgate of her pickup, spread with a clean burlap sack; her scalpels, syringes, and other tools came from a portable kit. First she anesthetized Noddy with gas. Then, after drawing blood samples from deep in his heart, she injected him with something called Lethabarb, which killed him. She measured his carcass, inspected his face, and then sliced an olive-size lump off his right cheek just below the eye. She showed me the lump's interior: a pea-size core of pale tissue surrounded by normal pink flesh. She put a chunk of it into a vial; that would go to a lab up at Mount Pleasant, she said, to be grown for chromosome typing. From the left side of the face, among the whiskers, Harmsen cut another tumor. Noddy lay limp on the burlap, both cheeks sliced away, like a halibut. When she slit open his belly and found an abundance of healthy yellow fat, she sighed. "He was in good condition." The disease hadn't progressed far. There was no sign of metastasis. But the protocols of the trapping program on Forestier don't include therapeutic surgery and chemo. Harmsen put a bit of Noddy's liver, a bit of his spleen, and a bit of his kidney into formalin. Those samples, too, would go back to Mount Pleasant for analysis. She wrapped the rest of Noddy in his burlap shroud, put him in a plastic garbage bag, and sealed that with tape. He would be incinerated. Then she cleaned up, fastidiously, to eliminate the chance that tumor cells might pass from her tailgate or her tools to another animal.

THE PHENOMENON OF TRANSMISSIBLE TUMORS isn't confined to canines, Tasmanian devils, and Syrian hamsters. There have been human cases, too. Forty years ago a team of physicians led by Edward F. Scanlon reported, in the journal *Cancer*, that they had "decided to transplant small pieces of tumor from a cancer patient into a healthy donor, on a well informed volunteer basis, in the hope of gaining a little better understanding of cancer immunity," which they thought might help in treating the patient. The patient was a fifty-year-old woman with advanced melanoma; the "donor" was her healthy eighty-year-old mother, who had agreed to receive a bit of the tumor by surgical transplant. One day after the transplant procedure, the daughter died suddenly from a perforated bowel. Scanlon's report neglects to explain why the experiment wasn't promptly terminated—why they didn't dive back in surgically to undo what had been done to the mother. Instead, three weeks were allowed to pass, at which point the mother had developed a tumor indistinguishable from her daughter's. Now it was too late for surgery. This cancer moved fast. It metastasized, and the mother died about fifteen months later, with tumors in her lungs, ribs, lymph nodes, and diaphragm.

The case of the daughter-mother transplant and the case of the Syrian hamsters have one common element: the original sources of the tumor and the recipients were genetically very similar. If the genome of one individual closely resembles the genome of another (as children resemble their parents, and as inbred animals resemble one another), the immune system of a recipient may not detect the foreignness of transplanted cells. The hamsters were highly inbred (intentionally, for experimental control) and therefore not very individuated from one another as far as their immune systems could discern. The mother and daughter were also genetically similar—as similar as two people can be without being identical twins. Lack of normal immune response, because of such closeness, goes some way toward explaining why those tumors survived transference between individuals.

Low immune response also figures in two other situations in which tumor transmission is known to occur: pregnancy and organ transplant. A mother sometimes passes cancer cells to her fetus in the womb. And a transplanted organ sometimes carries tiny tumors into the recipient, vitiating the benefits of receiving a life-saving liver or kidney from someone else. Cases of both kinds are very rare, and they involve some inherent or arranged compatibility between the original victim of the tumor and the secondary victim, plus an immune system that is either compromised (by immuno-suppressive drugs, in the organ recipient) or immature (in the fetus).

Other cases are less easily explained. In 1986, two researchers from the National Institutes of Health reported that a laboratory worker, a healthy nineteen-year-old woman, had accidentally jabbed herself with a syringe carrying colon-cancer cells; a colonic tumor grew in her hand, but she was rescued by surgery. More recently, a fifty-three-year-old surgeon cut his left palm while removing a malignancy from a patient's abdomen, and five months later he found himself with a palm tumor, one that genetically matched the patient's tumor. His immune system responded, creating an inflammation around the tumor, but the response was insufficient and the tumor kept growing. Why? How? It wasn't supposed to be able to do that. Again, though, surgery delivered a full cure. And then there's Henri Vadon. He was a medical student in the 1920s who poked his left hand with a syringe after drawing liquid from the mastectomy wound of a woman being treated for breast cancer. Vadon, too, developed a hand tumor. Three years later, he died of metastasized cancer because neither the surgical techniques of his era nor his own immune system could save him.

THE TUMOR THAT I HAD watched Colette Harmsen harvest from Noddy's face would be examined at the Mount Pleasant labs by Anne-Maree Pearse and her assistant Kate Swift. Pearse is a former parasitologist, now working in cell biology, and she has a special interest in the genetics of Devil Facial Tumor Disease. She and Swift were the researchers who, in 2006, published a dramatic report in the British journal *Nature* that, with eight paragraphs of text and a single photographic image, had answered the lingering question about whether DFTD is a genuinely transmissible cancer.

Pearse came out of retirement (she had turned to running a flower farm) in response to the scientific conundrum of DFTD. A back injury has forced the use of a cane on her, but she is vigorous when describing her research. Although she was originally trained as an entomologist, her work with fleas drew her into the world of parasitology, and from there it was just a few more steps into oncology and the study of lymphomas among the devil and its close marsupial relatives. "Somehow my whole life was preparing me for this," she said when I visited her lab at Mount Pleasant. She added, almost appreciatively: "This disease." Pearse tends to think, as she put it, "outside the square"—a useful trait in the case of DFTD, she said, because the disease isn't behaving like anything heretofore known. "It's a parasitic cancer," she told me. "The devil's the host."

For the 2006 study, Pearse and Swift examined chromosome structure in tumor cells from eleven different devils. They found that the tumor chromosomes were abnormal (misshapen, some missing, some added) compared with those from healthy devil cells, but that the tumor chromosomes, from one cell or another, from one tumor or another, were abnormal in *all the same ways*. You could see that comparison graphically in the photo in *Nature*: fourteen nice sausages matched against thirteen variously mangled ones. Those thirteen chromosomes, wrote Pearse and Swift, had undergone "a complex rearrangement that is identical for every animal studied." The mangling was unmistakable evidence. It appeared in each tumor, but not elsewhere in each animal. "In light of this remarkable finding and of the known fighting behavior of the devils," Pearse and Swift wrote, "we propose that the disease is transmitted by allograft"—tissue transplant—"whereby an infectious cell line is passed directly between the animals through bites they inflict on one another."

Pearse and Swift had proved that DFTD is a highly infectious form of cancer, its transmission made possible by, among other factors, the habit of mutual face-biting. When I visited Menna Jones, she expressed the same idea: "It's a piece of devil tissue that behaves like a parasite." Jones was using the word "parasite" in its strict biological sense, meaning: any organism that lives on or within another kind of organism, extracting benefit for itself and causing harm to the other. The first rule of a successful parasite is, Don't kill your host—or at least, Don't kill one host until you've had time to leap aboard another. DFTD, passing quickly from devil to devil, killing them all but not quite so quickly, follows that rule.

HOW DOES ANY PARASITE, WHETHER it is a species or merely a tumor, acquire the attributes and tactics necessary for survival, reproduction, and continuing success? The answer is simple but not obvious: evolution.

Cancer and evolution have traditionally been considered separately by different scientists with different interests using different methods. You could graduate from medical school, you could follow that with a Ph.D. in cell biology or molecular genetics, you could become a respected oncologist or a well-funded cancer researcher, without ever having read Darwin. You could do it, in fact, without having studied much evolutionary biology at all. Many cell and molecular biologists tended even to scorn evolutionary biology as a "merely descriptive" enterprise, lacking the rigor, quantifiability, and explanatory power of their disciplines. There were exceptions to this disconnect, cancer scientists who even during the early days thought in evolutionary terms, but those scientists had little influence.

In recent decades, however, the situation has changed, as molecular genetics and evolutionary biology have converged on some shared questions. One signal act of synthesis occurred in 1976 when a leukemia researcher named Peter Nowell published a theoretical paper in *Science* titled "The Clonal Evolution of Tumor Cell Populations." Nowell proposed what was then a novel idea: that the biological events occurring when cells progress from normal to pre-cancerous to cancerous represent a form of evolution by natural selection. As with the evolution of species, he suggested, the evolution of malignant tumors requires two conditions: genetic diversity among the individuals of a population and competition among those individuals for limited resources. Genetic diversity within one mass of precancerous cells comes from mutations—copying errors and other forms of change—that yield variants as the cells reproduce. That is, in the very act of replicating themselves (sometimes inaccurately), the cells diversify into a population encompassing some small genetic differences between one cell and another. Each variant cell then replicates itself true to type, constituting a clonal lineage (a lineage of accurate copies), until the next mutation creates a new variant. The fittest variants survive and proliferate. By this means, the genetic character of the cell population gradually changes, and with such change comes adaptation, a better fit to environmental circumstances. What constitutes "the fittest" among clonal lineages within a pre-cancerous growth? Those that can reproduce fastest. Those that can resist chemotherapy. Those that can metastasize and therefore escape the surgeon's knife.

Nowell's hypothesis about tumor evolution became widely known and accepted within certain circles of cancer research. (Among other researchers, it wasn't adamantly disputed but merely ignored.) Those circles have more recently produced a lot of rich theorizing, and a smaller amount of empirical work, supporting Nowell and carrying his idea forward. A culmination of sorts occurred in 2000, when the cancer geneticist Robert Weinberg, discoverer of the first human oncogene and the first tumor suppressor gene, published a concise paper titled "The Hallmarks of Cancer." Weinberg and his coauthor, Douglas Hanahan, described six "acquired capabilities," such as endless self-replication, the ignoring of antigrowth signals, the invasion of neighboring tissues, and the refusal to die, that collectively characterize cancer cells. How are those capabilities acquired? By mutations and other genetic changes, giving cells with one such trait or another competitive advantage over normal cells. Hanahan and Weinberg added that "tumor development proceeds via a process formally analogous to Darwinian evolution." With this cautious phrasing, they gave authoritative endorsement to the idea that Peter Nowell had proposed: Cancers, like species, evolve.

IN 1998 A YOUNG RESEARCHER named Carlo Maley began looking for a way to study the evolution of cancer. Educated at Oxford and MIT as an evolutionary biologist and a computer modeler, Maley had no training in medicine and not much in molecular biology. During a postgraduate fellowship, though, he became interested in infectious disease. He figured that if evolution was cool, then coevolution—wherein both parasite and host are evolving—would be doubly cool. Then he stumbled across a description of cancer as an evolving disease. He read that Sir Walter Bodmer, a British geneticist and the former director of the Imperial Cancer Research Fund, had urged his cancer-research colleagues to "think evolution, evolution, evolution" when they considered tumor cells. Maley typed "evolution and cancer" into a search engine for the scientific literature, which turned up very little. He did learn of Nowell's hypothesis, but that was just theory. He was groping. He had done plenty of theoretical modeling, but for this task he needed the desperate realities, and the data, of clinical oncology.

And then, at a workshop in Seattle, Maley met Brian Reid, an experienced cancer researcher studying something called Barrett's esophagus, a pre-cancerous condition of the lower throat. They hit it off. Reid had the right clinical situation but wasn't deeply versed in evolutionary biology; Maley had the right background. They agreed to collaborate.

Reid and his colleagues possessed sixteen years of continuous data on Barrett's patients and a tissue bank going back to 1989. They knew which patients had developed esophageal cancer and which hadn't, and they could match those outcomes against what they had seen in cell cultures and genetic work from earlier in the patients' history. So they could ask evolutionary questions that were answerable from patterns in the data. The most basic question was: Did tumors become malignant through evolution by natural selection? The other big question was: Can doctors predict which pre-cancerous growths will turn malignant? Maley and Reid, along with additional collaborators, found that case histories of Barrett's esophagus tend to confirm Nowell's hypothesis. Cancerous tumors, like species, *do* evolve. And from the Barrett's data, predictions can be made. The higher the diversity of different cell variants within a pre-cancerous growth, the greater the likelihood that the growth will progress to malignancy. Why? Because of the basic Darwinian mechanism. Genetic diversity plus competitive struggle eliminates unfit individuals and leaves the well-adapted to reproduce.

Maley and Reid have more recently taken such thinking one step beyond evolution—into ecology. Along with Lauren M. F. Merlo (as first author) and John W. Pepper, they published a provocative paper titled "Cancer as an Evolutionary and Ecological Process," in which they discussed not just tumor evolution but also the ecological factors that form evolution's context, such as predation, parasitism, competition, dispersal, and colonization. Dispersal is travel by venturesome individuals, which in some cases allows species to colonize new habitats. Merlo, Maley, and their colleagues noted three ways in which the concept of dispersal is applicable to cancer: small-scale cell movement within a tumor (not very important), invasion of neighboring tissues (important), and metastasis (fateful).

Reading that, I remembered Devil Facial Tumor Disease and wondered whether there might not be a fourth way: transmissibility. An infectious cancer is a successful disperser. It colonizes new habitat. DFTD seems to be dispersing and colonizing, much as pigeons disperse across oceans, colonizing new islands. This wasn't just evolution; it was evolutionary ecology.

I called one of the paper's coauthors, John W. Pepper, an evolutionary biologist at the University of Arizona, and asked whether I was stretching the notion too far. No, he said, you're not. If he could revise that paper again, Pepper told me, he would insert the idea that tumors evolve toward transmissibility.

EIGHT HUNDRED MILLION YEARS AGO there was no such thing as cancer. Virtually all living creatures were single-celled organisms, and the rule was *Every cell for itself!* Uncontrolled, undifferentiated cell growth wasn't abnormal. It was the program of all life on Earth.

Then, around 700 million years ago, things changed. Paleontologists call this event the Cambrian explosion. Complex multicellular animals, metazoans, appeared. And not just meta*zoans* but meta*phytes*, too—that is, multicellular plants. How did it happen? Very gradually, as single-cell creatures resembling bacteria or algae began to aggregate into colonial units and discover, by trial and error, how they could benefit from division of labor and specialization of shape and function. To enjoy those benefits, they had to set aside the old rule of absolute selfishness. They had to cooperate. They couldn't cheat against the interests of the collective entity. (Or anyway they *shouldn't* cheat, not very often; otherwise the benefits of collectivity wouldn't accrue.) Cooperation was a winning formula. Primitive multicellular creatures, roughly along the lines of jellyfish or sponges or slime molds, began to succeed, to grow, to occupy space, and to claim resources in ways that loner cells couldn't. You can see their imprints in the Burgess Shale: weird things like sci-fi vermin, pre-vertebrate, pre-insect, that seem to have been built out of bubble wrap and old Slinkys. They succeeded for a while, then gave way to still better designs. Multicellularity offered wide possibilities.

But uncontrolled cell replication didn't disappear entirely. Sometimes a single atavistic cell would ignore the collective imperative; it would revert to the old habit—proliferating wildly, disregarding all signals to stop. It would swell into a big, greedy lump of its own kind, and in so doing disrupt one or more of the necessary collective functions. That was cancer.

The risk of runaway cell replication remained a factor in the evolutionary process, even as multicellular creatures increased vastly in complexity, diversity, size, and dominance on our planet. And species responded to that risk just as they responded, incrementally and over long periods of time, to other such risks as predation or parasitism: by acquiring defenses. One such defense is the amazing ability of living cells to repair mutated DNA, putting the cell program back together properly after a mishap during cell replication. Another defense is *apoptosis*, a form of programming that tells a cell not to live forever. Another is cellular senescence, during which a cell continues to live but no longer is capable of replicating. Another is the distinction between stem cells and differentiated cells, which limits the number of cells responsible for cell-replacement activity and thereby reduces the risk of accumulated mutations. Another is the requirement for biochemical growth signals before a cell can begin to proliferate. Many of these defenses are controlled by tumor-suppressor genes, such as the one that produces a protein that prevents cells from replicating damaged DNA. Nobody knows just how many anticancer defenses exist within a given species (we humans seem to have more than mice do, and possibly not so many as whales), but we do know that they make our continuing lives possible.

Tumors, in the course of their own evolution from one normal cell to a cancerous malignancy, circumvent these natural defenses. They may also change in response to externally imposed defenses, such as surgery, chemotherapy, and radiation. The fittest cells, in Darwinian terms, are those that reproduce themselves most quickly and aggressively, resisting all signals to desist and all attempts to kill them. The victim (that is, the human or the Tasmanian devil or the Syrian hamster) suffers the consequences, having become the arena for an evolutionary struggle at a scale far different from that of its own struggle to survive. But the principles of the struggle are the same at each scale.

This process, whereby cells mutate, reproduce, and proliferate differentially within a body, is called somatic evolution. It stands as a counterpoint to organismic evolution (progressive changes at the scale of whole bodies within a population), and the opportunities for it to proceed are abundant. According to one count, at least 291 genes in the human body contribute, when damaged by mutation, toward somatic evolution.

Mutations occur when something goes wrong during cell replication. A cell replicates by copying its DNA (sometimes inaccurately), sorting the DNA into two identical parcels of chromosomes, then splitting into two new cells, each with its own chromosomes. This process is called mitosis. The goal is not to generate an ever-higher number of cells during a creature's adult lifetime but simply to replace old cells with new ones. Mitosis counterbalanced with apoptosis, cell death, should provide a constant supply. But each time a mitotic division occurs, there is some very small chance of mutation. And the many small chances add up. A human body contains about 30 trillion cells. The number of cell divisions that occur in a lifetime is far larger: 10,000 trillion. A disproportionately large share of those divisions occurs in epithelial cells, which serve as boundary layers or linings, such as the skin and the interior surface of the colon. That's why skin cancer and colon cancer are relatively common—more cell turnover, more chance of mutation and evolution.

How many mutations does it take for a malignancy to occur? Estimates range from three to twelve, in humans, depending on the form of cancer. Five or six is considered an operational average for purposes of discussion. Here's some good news: For a cell to acquire those five or six changes, at the usual rate of mutation, is highly, highly, highly unlikely. The odds are great against quintuple mutation in any given cell, making cancer seem impossible within a human lifespan. One form of mutation, however, can vastly increase the later rate of mutation, which gives the pre-cancerous cells many more chances to become malignant.

In the United States, about 40 percent of us will eventually get cancer bad enough to be diagnosed. And autopsies suggest that virtually all of us will be nurturing incipient thyroid cancer by the time we die. Among octogenarian and nonagenarian men, 80 percent carry prostate cancer when they go. Cancer is terrible, cancer is dramatic, but cancer isn't rare. In fact, it's nearly universal.

THE BIOLOGICAL MYSTERY OF HOW the Tasmanian devil's rogue tumor manages to establish itself in one animal after another is still unsolved. But a good hypothesis has been offered by an immunologist named Greg Woods at the University of Tasmania. Woods and his group studied immune reactions in *Sarcophilus harrisii*, which seem generally to be normal against ordinary sorts of infection. Against DFTD cells, though, no such reaction occurs. "The tumor is just not *seen* by the immune system, because it just looks too similar," Woods told me when I stopped by his lab. The devils have low genetic diversity, probably because they inhabit a small island, they colonized it originally by way of just a few founders, and they have passed through some tight population bottlenecks in the centuries since. They're not quite so alike one another as a bunch of inbred hamsters, but they're too alike for their own good in the current sad, anomalous circumstances. Their immune systems don't reject the tumor cells because, Woods suspects, in each animal the critical MHC genes (the major histocompatibility complex, which produces proteins crucial to immunological policing) are all virtually identical, and the devils' police cells can't distinguish "him" from "me."

Most of the DFTD team are, like Greg Woods and Anne-Maree Pearse, located in Hobart or Launceston. Most of the animals aren't. So, two days after the autopsy on Noddy, I drove back to the Forestier Peninsula for another round of trapping with Chrissy Pukk and her crew. It wasn't that I expected to learn any new angles on the science. I just wanted to see more Tasmanian devils.

This time, Pukk issued me my own pair of blue coveralls. As we set off along the trap line, she exuded the contentment of a joyously crude tomboy enjoying the best job in the world: trapping devils for the good of the species. The only downside was the necessity of issuing a death sentence to any animal with a trace of tumor. "You get attached to the individuals," she admitted. "But you've got to remember all the other individuals you can save if you take that animal out early on."

Wouldn't this be less difficult emotionally, I asked her, if you gave them *numbers* instead of names? She answered the question—saying she couldn't do her work properly if she wasn't emotionally invested, plus which, names were easier to remember—and then she continued to answer it throughout the day. These creatures, they all have their memorable eccentricities, their little histories, she explained. Some she could recognize almost by smell. You couldn't do *that* with numbers.

There were forty traps again today, and about a dozen trapped devils to process, all recaptures, previously tagged and named. Trap by trap, animal by animal, Pukk worked through the measurements and the facial exams, handling each devil firmly but with a steady touch that provoked no devilish squirming: Captain Bligh (looking glum, or maybe a little embarrassed at having been caught again so soon), Hipster, Isabel, Masikus (Estonian for "strawberry"), Miss Buzzy Bum (her rump had been full of burrs), Rudolph (thus called for a nose that had been rubbed raw), Sandman, Skipper, and many others. They may have been virtually indistinguishable in the terms by which immune systems operate, but Chrissy Pukk knew each devil at a glance.

Rudolph's condition gave her pause. He was a two-year-old, nicely grown since she had first trapped him, his red nose healed . . . but there was something on the edge of his right eye. A pink growth, no bigger than a caper. "Oh shit," she said. Tumor? Or maybe it was just a little wound, puffy and raw. She looked closely. She peered into his mouth. She palpated lymph nodes at the base of his jaw. The volunteers and I waited in silence. Evolution had shaped Rudolph for survival, and evolution might take him away. It was all evolution: the yin of struggle and death, the yang of adaptation, DFTD versus *Sarcophilus harrisii*. The leaping tumor, well adapted for fast replication and transmissibility, has its own formidable impulse to survive. And no one could know at this point, not even Chrissy, whether it had already leapt into Rudolph.

"Okay," she said, sounding almost sure of her judgment, "I'm gonna give him the all clear." She released him and he ran.

At latest report, Devil Facial Tumor Disease has spread across 60 percent of Tasmania's land surface, and in some areas, especially where it got its earliest start, the devil population seems to have declined by as much as 90 percent. In November, the Tasmanian government classified the devil as "endangered." DFTD specialists differ strongly on how such a crisis should be met. One view is that suppressing the disease—trapping and euthanizing as many infected animals as possible and then establishing barriers, as on the Forestier Peninsula—is the best strategy. Another view is that the species, virtually doomed on mainland Tasmania, can better be saved by transplanting disease-free devils to a small offshore island. Still another view, maybe the boldest and most risky, is that doing nothing—allowing the disease to spread unchecked—might yield a small remnant population of survivors, with natural immunity to DFTD, who could repopulate Tasmania.

Weeks after my last outing with the trappers, back in the Northern Hemisphere and wanting a broader perspective—not just on the fate of the devils but on the evolution of cancer—I met Robert Weinberg at a stem-cell conference in Big Sky, Montana. Because it was a Sunday, with the first session not yet convened, and because he is a genial man, Weinberg gave me a two-hour tutorial in a boardroom of the ski lodge, fortifying some points by flipping through his own four-pound textbook, *The Biology of Cancer*, a copy of which I had lugged to our meeting like a student. He was incognito in a plaid Woolrich shirt. He'd be called on that evening to deliver the keynote address, but never mind, he was prepared.

"Infectious cancer is really an aberration," Weinberg told me, affirming what Greg Woods had said. "It's so bizarre. It has happened only rarely." Maybe it's possible only in cases where there's close physical contact between susceptible tissues. "That, right away, limits it to venereal tumors, or tumors that can be transmitted by biting." Weinberg knew that I'd walked in with a head full of Tasmanian devils.

Does this mean that cancer cells are harder to transmit than, say, virus particles? "Much," he said. "Cells are very *effete.* Very susceptible to dying in the outside world." They dry out, they wither, they don't remain viable when they're naked and alone. Bacteria can form spores. Viruses in their capsules can lie dormant. But cells from a metazoan? No. They're not packaged for transit.

And that's only one of two major constraints, Weinberg said. The second is that if cancer cells *do* pass from one body to another, they are instantly recognized as foreign and eliminated by the immune system. Each cell of any sort bears on its exterior a set of protuberant proteins that declare its identity; they might be thought of as its travel papers. These proteins are called antigens and are produced uniquely in each individual by the MHC (major histocompatibility complex) genes. If the travel papers of a cell are unacceptable (because the cell is an invader from some other body), the T cells (one type of immunological police cell) will attack and obliterate it. If the invader cell shows no papers at all, another kind of police cell (called NK cells) will bust it. Only if the antigens on the cell surface have been "downregulated" discreetly but not eliminated altogether can a foreign cell elude the immune system of a host. That's what Canine Transmissible Venereal Tumor seems to have done: downregulated its antigens. It shows fake travel papers—blurry, faded, but just good enough to get by.

Nice trick! How did CTVT do that? Although nobody knows exactly, the best hypothesis is evolution by natural selection—or by some process "formally analogous" to it.

Weinberg went on to explain that the process is a little more complicated than classic Darwinian selection. Darwin's version works by selection among genetic variations that differentiate one organism from another, and in sexually reproducing species those variations are heritable. But evolution in tumor lineages occurs by that sort of selection plus another sort—selection among *epigenetic* modifications of DNA. Epigenetic means outside the line of genetic inheritance: acquired by experience, by accident, by circumstance. Such secondary chemical changes to the molecule affect behavior, affect shape, and pass from one cell to another but do *not*, contrary to the analogy, pass from parent to offspring in sexual reproduction. These changes are peeled away in the process of meiosis (the formation of sperm and egg cells for sexual reproduction) but preserved in mitosis (the process of simple cell replication in the body). So cancerous cell reproduction brings such changes forward into the new cells, along with the fundamental genetic changes.

Does that mean tumors don't evolve? Certainly not. They do. "It's still Darwin," Weinberg said. "It's Darwin revised."

JENNIFER MARGULIS

Looking Up

FROM *SMITHSONIAN*

In Niger, giraffes are part of the landscape—perhaps too much so, as villagers compete with these majestic animals for precious resources. Jennifer Margulis describes what is being done to mitigate this conflict.

IN THE DRY SEASON, THEY ARE HARD TO FIND. FOOD IS scarce in Niger's bush and the animals are on the move, loping miles a day to eat the tops of acacia and combretum trees. I'm in the back seat of a Land Rover and two guides are sitting on the roof. We're looking for some of the only giraffes in the world that roam entirely in unprotected habitat.

Though it's well over 90 degrees Fahrenheit by 10 a.m., the guides find it chilly and are wearing parkas, and one of them, Kimba Idé, has pulled a blue woolen toque over his ears. Idé bangs on the windshield with a long stick to direct the driver: left, right, right again. Frantic tapping means slow down. Pointing into the air means speed up. But it's hard to imagine going any faster. We are off-road, and the bumps pitch us so high that my seat belt cuts into my neck and my tape recorder flies into the front seat, prompting the driver to laugh. Thorny bushes scraping the truck's paint sound like fingernails on a chalkboard. I don't know what to worry about more: the damage the truck might be causing to the ecosystem or the very real possibility we might flip over.

While Africa may have as many as 100,000 giraffes, most of them live in wildlife reserves, private sanctuaries, national parks or other protected areas not inhabited by humans. Niger's giraffes, however, live alongside villagers, most of whom are subsistence farmers from the Zarma ethnic group. Nomadic Peuls, another group, also pass through the area herding cattle. The "giraffe zone," where the animals spend most of their time, is about 40 square miles, although their full range is about 650 square miles. I've seen villagers cutting millet, oblivious to giraffes foraging nearby—a picturesque tableau. But Niger is one of the poorest, most desolate places on earth—it has consistently ranked at or near the bottom of the 177 nations on the United Nations Human Development Index—and people and giraffes are both fighting for survival, competing for some of the same scarce resources in this dry, increasingly deforested land.

There are nine giraffe subspecies, each distinguished by its range and the color and pattern of its coat. The endangered *Giraffa camelopardalis peralta* is the one found in Niger and only Niger; it has large orange-brown spots on its body that fade to white on its legs. (The reticulated subspecies, known for its sharply defined chestnut brown spots, is found in many zoos.) In the 19th century, thousands of peralta giraffes lived in West Africa, from Mauritania to Niger, in the semiarid land known as the Sahel. By 1996, fewer than 50 remained because of hunting, deforestation and development; the subspecies was heading for extinction.

That was about the time I first went to Niger, to work for a development organization called Africare/Niger in the capital city of Niamey. I recall being struck by the heartbreaking beauty of the desert, the way people managed to live with so little—they imported used tires from Germany, drove on them until they were bald and then used them as soles for their shoes—and the slower pace of life. We drank mint tea loaded with sugar and sat for hours waiting for painted henna designs to dry on our skin. "I don't know how anyone can visit West Africa and want to live anywhere else in the world," I wrote in my journal as an idealistic 23-year-old.

Two nights a week I taught English at the American Culture Center, where one of my students was a young French ethologist named Isabelle Ciofolo. She spent her days following the giraffes to observe their behavior. She would study the herd for 12 years and was the first to publish research about it. In 1994, she helped found the Association to Safeguard the Giraffes of Niger (ASGN), which protects giraffe habitat, educates the local population about giraffes, and provides microloans and other aid to villagers in the giraffe zone. The ASGN also participates in an annual giraffe census. Which is how I ended up, some 15 years after I first met Ciofolo, in a bucking Land Rover on a giraffe observation expedition that she was leading with Omer Dovi, the Nigerien operations manager for ASGN.

Working on a tip that a large group of giraffes had been spotted the night before, we spend more than two hours looking for them in the bush before we veer off into the savanna. Another hour goes by before Dovi shouts, "There they are!" The driver cuts the Land Rover's engine and we approach the animals on foot: a towering male with large brown spots, two females and three nurslings, which are all ambling through the bush.

The adult giraffes pause and regard us nonchalantly before going back to their browsing. The nurslings, which are only a few weeks old and as frisky as colts, stop and stare at us, batting enormous Mae West eyelashes. Their petal-shaped ears are cocked forward beside their furry horns (which, Ciofolo says, are not really horns but ossicones made from cartilage and covered with skin). Not even the guides can tell if the nurslings are male or female. Once a giraffe matures, the distinction is easy: peralta males grow a third ossicone. The census-takers note three baby giraffes of indeterminate gender.

We watch the statuesque animals galumph forward in the bush. They are affectionate, intertwining necks and walking so closely that their flanks touch. They seem in constant physical contact, and I am struck by how much they seem to enjoy each other's presence.

I ask Ciofolo if she thinks giraffes are intelligent. "I'm not sure how to evaluate the intelligence of a giraffe," she says. "They engage in subtle communication with each other"—grunts, snorts, whistles, bleats—"and we have observed that they are able to figure things out." Ciofolo says a giraffe she named Penelope years ago (the scientists now designate individual animals less personally, with numbers) "clearly knew who I was and had assessed that I was not a threat to her. She let me get quite close to her. But when other people approached, she got skittish. Penelope was able to distinguish perfectly between a person who was nonthreatening and people who represented a potential threat."

A year later, in late 2007, I return to Niger and go into the bush with Jean-Patrick Suraud, a doctoral student from the University of Lyon and an ASGN adviser, to observe another census. It takes us only half an hour to find a cluster of seven giraffes. Suraud points out a male that is closely following a female. The giraffe nuzzles her genitals, which prompts her to urinate. He bends his long neck and catches some urine on his muzzle, then raises his head and twists his long black tongue, baring his teeth. Male giraffes, like snakes, elephants and some other animals, have a sensory organ in their mouth, called Jacobson's organ, that enables them to tell if a female is fertile from the taste of her urine. "It's very practical," Suraud says with a laugh. "You don't have to take her out to dinner, you don't have to buy her flowers."

Although the female pauses to let the male test her, she walks away. He does not follow. Presumably she is not fertile. He meanders off to browse.

If a female is fertile, the male will try to mount her. The female may keep walking, causing the male's forelegs to fall awkwardly back to the ground. In the only successful coupling Suraud has witnessed, a male pursued a female—walking alongside her, rubbing her neck, swaying his long body to get her attention—for more than three hours before she finally accepted him. The act itself was over in less than ten seconds.

Suraud is the only scientist known to have witnessed a peralta giraffe give birth. In 2005, after just six months in the field, he was stunned when he came upon a female giraffe with two hoofs sticking out of her vagina. "The giraffe gave birth standing up," he recalls. "The calf fell [six feet] to the ground and rolled a bit." Suraud smacks the top of the truck to illustrate the force of the landing. "I'd read about it before, but still, the fall was brutal. I remember thinking, 'Ouch, that's a crazy way to come into the world.'" The fall, he goes on, "cuts the umbilical cord in one swift motion." Suraud then watched the mother lick the calf and eat part of the placenta. Less than an hour later, the calf had nursed and the two were on the move.

Though mother and calf stay together, groups of giraffes are constantly forming and re-forming in a process scientists call fission-fusion, similar to chimpanzee grouping. It is as common for half a dozen males to forage together as it is for three females and a male. In the rainy season, when food is plentiful, you might find a herd of 20 or more giraffes.

Unlike with chimps, however, it is almost impossible to identify an alpha male among giraffes. Still, Suraud says he has seen male giraffes mount other males in mock copulation, often after a fight. He's not sure what to make of the behavior but suggests it may be a type of dominance display, though there doesn't seem to be an overarching power hierarchy. Competition among males—which grow to 18 feet tall and weigh as much as 3,000 pounds—for access to females, which are slightly smaller, can be fierce. Males sometimes slam each other with their necks. Seen from afar, a fight might look balletic, but the blows can be brutal. Idé says he witnessed a fight several years ago in which the vanquished giraffe bled to death.

As it happens, the evolution of the animal's neck is a matter of some debate. Charles Darwin wrote in *The Origin of Species* that the giraffe is "beautifully adapted for browsing on the higher branches of trees." But some biologists suggest that the emergence of the distinctive trait was driven more by sexual success: males with longer necks won more battles, mated more often and passed on the advantage to future generations.

STILL, WILD GIRAFFES NEED A lot of trees.

They live up to 25 years and eat from 75 to 165 pounds of leaves per day. During the dry season, Niger's giraffes get most of their water from leaves and the morning dew. They're a bit like camels. "If water is available, they drink and drink and drink," says Suraud. "But, in fact, they seem not to have a need for it."

Dovi points out places in the savanna where villagers have cut down trees. "The problem is not that they take wood for their own use; there's enough for that," he says. "The problem is that they cut down trees to sell to the market in Niamey."

Most woodcutting is prohibited in the giraffe zone. But Lt. Col. Kimba Ousseini, commander of the Nigerien government's Environmental Protection Brigade, says people break the law, despite penalties of between 20,000 and 300,000 CFA francs (approximately $40 to $600) as well as imprisonment. He estimates that 10 to 15 people are fined each year. Yet wood is used to heat houses and fuel cook-fires, and stacks and stacks of spindly branches are for sale at the side of the road to Niamey.

When you walk alongside the towering giraffes, close enough to hear the swish-swish of their tails as they gambol past, it's hard not to be indignant about the destruction of their habitat. But Zarma villagers cut down trees because they have few other ways to make money. They live off their crops and are totally dependent on the rainy season to irrigate their millet fields. "Of course they understand why they shouldn't do it!" Ousseini says. "But they tell us they need the money to survive."

The ASGN is trying to help the giraffes by making small loans to villagers and promoting tourism and other initiatives. In the village of Kanaré, women gathered near a well constructed with ASGN funds. By bringing aid to the region in the name of protecting giraffes, ASGN hopes the villagers will see the animals as less of a threat to their livelihood. A woman named Amina, who has six children and was sitting in the shade on a wire-and-metal chair, says she benefited from an ASGN microloan that enabled her to buy goats and sheep, which she fattened and sold. "Giraffes have brought happiness here," Amina says in Zarma through an interpreter. "Their presence brings us lots of things."

At the same time, giraffes can be a nuisance. They occasionally eat crops such as niebe beans, which look like black-eyed peas and are crushed into flour. (We ate tasty niebe-flour beignets for breakfast in a village called Harikanassou, where we spent the night on thin mattresses under mosquito nets.) Giraffes splay their legs and bend their long necks to eat mature beans right before harvest. They also forage on the succulent orange mangoes that ripen temptingly at giraffe-eye height.

The villagers' feelings about the giraffes, from what I gather after speaking with them, are not unlike what people in my small town in southern Oregon feel about deer and elk: they admire the animals from a distance but turn against them if they raid their gardens. "If we leave our niebe in the fields, the giraffes will eat it," explains Ali Hama, the village chief of Yedo. "We've had problems with that. So now we harvest it and bring it into the village to keep it away from the giraffes." Despite having to do this extra step, Hama says his villagers appreciate the giraffes because the animals have brought development to the region.

Unlike giraffes in other parts of Africa, Niger's giraffes have no animal predators. But they face other dangers. During the rainy season, giraffes often come to the Kollo road, about 40 miles east of Niamey, to nibble on shrubs that spring from the hard orange earth. On two occasions in 2006, a bush taxi hit and killed a giraffe at dusk. No people were injured, but the deaths were a significant loss to the small animal population. Villagers feasted on the one-ton animals.

The Niger government outlaws the killing of giraffes, and Col. Abdou Malam Issa, a Ministry of the Environment official, says the administration spends about $40,000 annually on anti-poaching enforcement. In addition, Niger has received money from environmental groups around the world to support the giraffes. As a result, giraffes face little danger of being killed as long as they stay within Niger. But when a group of seven peraltas strayed into Nigeria in 2007, government officials from Niger were unable to alert Nigerian officials quickly enough. Villagers killed one of the giraffes and ate it.

Niger's government hasn't always been disposed to help the giraffes. In 1996, after seizing power in a coup d'état, Ibrahim Baré Mainassara wanted to give two giraffes each to the presidents of Burkina Faso and Nigeria. When the forestry service refused to help him capture the giraffes, Baré sent in the army. More than 20 giraffes were killed, out of a total population of fewer than 60. "We lost 30 percent of the herd," says Ciofolo, who was working in the field at that time. In 2002, President Mamadou Tandja, who was first elected in 1999 and remains in power, set out to give a pair of giraffes to Togo's president. This time the Togolese Army, helped by local villagers and the forestry service, spent three days chasing the giraffes and captured two. One died en route to Togo, and the other after arriving there. Hama Noma, a 27-year-old villager who witnessed the capture, says the giraffes were immobilized with ropes and transported in the back of a truck: "They suffered a lot before they died."

* * *

DRIVING NORTH PAST A PITTED and rusty sign for the town of Niambere Bella, we come across a lone male strutting through the fields. "Number 208!" Suraud cries out. "This is only the second time I've seen him!" We find a group of 16 giraffes, an unusual sight during the dry season. Each one has been identified previously, which makes the research team rejoice. "It means we haven't missed any," says Suraud, clearly pleased. He pats Idé on the back, smiling. The mood is hopeful—at least 21 calves have been born recently, more than expected. And indeed the official results are heartening: 164 giraffes were photographed in 2007, leading the researchers to estimate that the population is around 175 individuals. While that number is dangerously small, it's up from 144 in 2006 and represents a 250 percent increase since 1996. Suraud says he is optimistic about the herd.

Julian Fennessy, a founding member of the International Giraffe Working Group of the International Union for Conservation of Nature, projects that a minimum of 400 giraffes of a variety of ages is needed for a viable peralta population. Whether the mostly desert climate of this part of West Africa can support the growing number remains to be seen; some giraffe researchers have even suggested that the giraffes might be better off in a wildlife refuge. But Ciofolo points out that the nearest reserve in Niger has unsuitable vegetation—and lions. "In my opinion, giraffes are much better off living where they are now, where they are protected by the local people," she says.

As the sky darkens, we drive past several villagers using handmade machetes called *coup-coups* to cut dried millet stalks. A father and son lead two bulls pulling a cart laden with straw bales along rough track in the bush. Now the royal blue sky is streaked with orange and violet from the setting sun, and the moon shimmers. Nearby, a group of foraging giraffes adds a calm majesty to the landscape these animals have so long inhabited.

MARGARET TALBOT

Birdbrain

FROM *THE NEW YORKER*

Alex, the African gray parrot who died in 2007, exhibited some remarkable traits—language skills and communication that went well beyond what one expects of parrots. For Alex's owner, Irene Pepperberg, Alex proved that animals are far more intelligent than we think. Other scientists are not convinced. Margaret Talbot tries to make sense of it all.

AS THE CROWD AT THE MIDWEST BIRD EXPO WAITED for the cognitive scientist Irene Pepperberg to take the podium, the hum of human chatter was punctuated by the sound of parrots whooping it up—twittering and letting loose with wolf whistles, along with the occasional full-out jungle squawk. The birds, many of them for sale, were displayed in cages just beyond the curtained-off stage, which was inside the main hall of the DuPage County Fairgrounds, in Wheaton, Illinois. Nobody seemed particularly distracted by the commotion. People were too busy pulling out their cell phones and showing one another photographs of their cockatiels back home. It was a warm Saturday afternoon in early April, and a woman in the folding metal chair in front of me, who was wearing large parrot earrings, said that she had driven all the way from Florida to see Pepperberg. Indeed, if this were a political rally, the audience would be Pepperberg's base. Here were admirers who had sent in ten-dollar bills to help support her research with Alex, the African gray parrot that she worked with for thirty years; and here were people who, after Alex died, unexpectedly, of heart arrhythmia, on September 6, 2007, helped form an online community that comes together on the sixth day of every month to reflect about him.

Pepperberg, arriving onstage, picked up a microphone and said, "I have a feeling you all know how smart Alex was." Everyone clapped in assent. (The parrot had appeared on television many times, mainly on PBS and the Discovery Channel.) She explained to the audience, which was largely middle-aged and female, that, in the late nineteen-seventies, when she started working with Alex—whose name was an acronym for Avian Learning Experiment—other scientists had been dismissive of her ambition to communicate with him. As she put it, "My grant proposals came back basically saying, 'What are you smoking?'" The woman from Florida laughed heartily, her parrot earrings bobbing.

Pepperberg, who is fifty-nine years old, has imposing cheekbones and an abundance of long, dark hair; she wears smoky eye makeup, short skirts, and an armful of silver bangles. In Wheaton, she quietly worked the crowd into a pleasurable state of shared outrage. At one point, she said that colleagues had admonished her, "Birds can't do what you say he can do. They just don't have the brainpower." Linnea Faris, a woman from Michigan who was wearing a "Remember Alex" T-shirt, shook her head in disbelief. Faris told me, "My husband doesn't really understand it. I can't fully explain it myself. But I've spent hours crying over that damn bird." She went on, "People used to think birds weren't intelligent. Well, they used to think women weren't intelligent, either. They talked about the smaller circumference of our skulls as though it made us inferior to men! You know what? They were wrong on both counts."

Pepperberg has had an unconventional academic career: she rents a small lab at Brandeis, and holds a part-time lecturer position at Harvard. That afternoon, she was delivering what she calls her "It's a Wonderful Life" speech, so named because it's about how surprised and touched she was to learn, after Alex died, that he had meant so much to people. "You all are my Clarences," she said, referring to the angel who shows Jimmy Stewart the sorry state of a Bedford Falls without him in it.

It wasn't just parrot people who had found themselves moved by Pepperberg's three-decade relationship with Alex: obituaries of the bird ran everywhere from *The Economist* to the *Hindustan Times.* Within a few days of his death, Pepperberg told the audience, she had received some six thousand messages of condolence via e-mail.

As everyone knows, parrots are remarkably good at mimicking human speech, but they tend to repeat randomly picked-up phrases: obscenities, election slogans, "Hey, sailor." Many parrots kept as pets also imitate familiar sounds, like the family dog barking or an alarm clock beeping. But Pepperberg taught Alex referential speech—labels for objects, and phrases like "Wanna go back." By the end, he knew about fifty words for objects. Pepperberg was never particularly interested in teaching Alex language for its own sake; rather, she was interested in what language could reveal about the workings of his mind. In learning to speak, Alex showed Pepperberg that he understood categories like same and different, bigger and smaller. He could count and recognize Arabic numerals up to six. He could identify objects by their color, shape ("three-corner," "four-corner," and so on, up to "six-corner"), and material: when Pepperberg held up, say, a pompom or a wooden block, he could answer "Wool" or

"Wood," correctly, about eighty per cent of the time. Holding up a yellow key and a green key of the same size, Pepperberg might ask Alex to identify a difference between them, and he'd say, "Color." When she held up two keys and asked, "Which is bigger?," he could identify the larger one by naming its color. Looking at a collection of objects that he hadn't seen before, Alex could reliably answer a two-tiered question like "How many blue blocks?"—a tricky task for toddlers. He even seemed to develop an understanding of absence, something akin to the concept of zero. If asked what the difference was between two identical blue keys, Alex learned to reply, "None." (He pronounced it "nuh.")

Pepperberg also reported that, outside training sessions, Alex sometimes played with the sounds he had learned, venturing new words. After he learned "gray," he came up with "grain" on his own, and after learning "talk" he tried out "chalk." His trainers then gave him the item that he had inadvertently named, and it eventually entered his vocabulary. (When Alex devised nonsense words—like "cheenut"—Pepperberg and his other trainers did not respond, and he quickly stopped saying them.) In linguistic terms, Alex was recombining phonemes, the building blocks of speech. Stephen Anderson, a Yale linguist who has written about animal communication, considers this behavior "apparent evidence that Alex did actually regard at least some of his words as made up of individual recombinable pieces, though it's hard to say without more evidence. This is something that seems well beyond any ape-language experiments, or anything we see in nature."

Pepperberg told me that Alex also made spontaneous remarks that were oddly appropriate. Once, when she rushed in the lab door, obviously harried, Alex said, "Calm down"—a phrase she had sometimes used with him. "What's your problem?" he sometimes demanded of a flustered trainer. When training sessions dragged on, Alex would say, "Wanna go back"—to his cage. More creatively, he'd sometimes announce, "I'm gonna go away now," and either turn his back to the person working with him or sidle as far away as he could get on his perch. "Say better," he chided the younger parrots that Pepperberg began training along with him. "You be good, see you tomorrow, I love you," he'd say when she left the lab each evening. This was endearing—and the *Times*' obituary made much of the fact that these were the bird's last words—although, as Anderson points out, it was during such moments that Alex was, most likely, merely "parroting." It helped Alex's charisma quotient that he made all his remarks in an intonation that was part two-year-old, part Rain Man, part pull-string toy. His voice, at once tinny and sweet, was easy to understand. Pepperberg tended to speak to Alex in the singsong "motherese" that doting parents use with young children, and he replied in a voice that seemed to convey a toddlerish pride.

Diana Reiss, a professor of cognitive psychology at Hunter College, is a dolphin researcher and a friend of Pepperberg's. (She was a co-author of the first studies to show that dolphins and elephants, like some primates, recognize themselves in a mirror.) Reiss recalls that when she first met Alex, "I was left alone with this bird I'd never seen. And I'm not really a bird person. Irene had said, 'Just do what you want with him.' After a few minutes, he said, 'Wanna go knee.' So I put out my hand, and he walked onto my knee. Then he said, 'Tickle,' and he put his head down to be tickled. I thought, I'm communicating with this bird! This is amazing! Then I thought, O.K., I'll test him. I knew Irene's paradigm. I knew she'd hold an object up really close to his beak. So I pick up a little red toy car and ask, 'What's this?,' and out he comes with 'Truck.' Now, if I were unconsciously cuing him with, say, subvocal speech—some subtle movement of my vocal cords or lips—I'd have said, 'Car.' But the label Alex had learned was 'truck.' The whole encounter knocked my socks off."

All children grow up in a world of talking animals. If they don't come to know them through fairy tales, Disney movies, or the Narnia books, they discover them some other way. A child will grant the gift of speech to the family dog, or to the stray cat that shows up at the door. At first, it's a solipsistic fantasy—the secret sharer you can tell your troubles to, or that only you understand. Later, it's rooted in a more philosophical curiosity, the longing to experience the ineffable interiority of some very different being. My eight-year-old daughter says that she wishes the horses she rides could talk, just so she could ask them what it feels like to be a horse. Such a desire presumes—as Thomas Nagel put it in his 1974 essay "What Is It Like to Be a Bat?"—that animals have some kind of subjectivity, and that it might somehow be plumbed. In any case, Nagel explained, humans are "restricted to the resources" of our own minds, and since "those resources are inadequate to the task," we cannot really imagine what it is like to be a bat, only, at best, what it is like to behave like one—to fly around in the dark, gobble up insects, and so on. That inability, however, should not lead us to dismiss the idea that animals "have experiences fully comparable in richness of detail to our own." We simply can't know. Yet many of us would be glad for even a few glimpses inside an animal's mind. And some people, like Irene Pepperberg, have dedicated their lives to documenting those glimpses.

Pepperberg will never really know what it is like to be a parrot, and she is careful not to claim that Alex learned a language. She calls what he learned a "two-way communication system," and prefers the more conservative term "labels," rather than "words," to describe his vocabulary. "I wouldn't say it was as much conversation as you'd have with a five-year-old," she said when I asked her about Alex's limits. "But maybe with a two-year-old. It wasn't like he was going to say, 'Hi, how are you, how was your day?' But if you came in and asked him what he wanted, where he wanted to go, and asked him questions within the context of labels he knew, you could talk to him."

At the Bird Expo, she told a story about the time an accountant was working on some tax forms near Alex's cage, and was more or less ignoring him. Peering down at the visitor, he asked her, "Wanna nut?" No, she said, not looking up. Want some water? No. A banana? No. And so on, through his repertoire of nameable desires. At last, Alex asked, in a tone in which it was hard not to detect a note of impatience, "What do you want?"

* * *

FOR CENTURIES, THE IDEA OF intelligent animals struck most intelligent people as ridiculous. In 1637, René Descartes marvelled that all humans—"without even excepting idiots"—can "arrange different words together, forming of them a statement by which they make known their thoughts, while there is no other animal, however perfect and fortunately circumstanced it may be, which can do the same." Two centuries later, Charles Darwin took a radically different view. The theory of evolution suggested that men and animals were separated not by an unbridgeable mental gulf but, rather, by developmental gradations. In *The Descent of Man*, Darwin contended that "there is no fundamental difference between man and the higher mammals in mental faculties." Chimps, he noted, used tools, and dogs had a capacity for abstraction—a poodle, seeing a German shepherd in the distance, knows it to be a fellow-canine. Discussing language, Darwin observed:

> That which distinguishes man from the lower animals is not the understanding of articulate sounds, for, as everyone knows, dogs understand many words and sentences. In this respect they are at the same stage of development as infants, between the ages of ten and twelve months, who understand many words and short sentences, but cannot yet utter a single word. It is not the mere articulation which is our distinguishing character, for parrots and other birds possess this power. Nor is it the mere capacity of connecting definite sounds with definite ideas; for it is certain that some parrots, which have been taught to speak, connect unerringly words with things, and persons with events. The lower animals differ from man solely in his almost infinitely larger power of associating together the most diversified sounds and ideas.

Darwin undermined these observations, however, with romantic overreaching about emotion in animals—he even made a claim for parental affection among earwigs. In 1881, Darwin's protégé George Romanes went still further; his book *Animal Intelligence* compiled hundreds of anthropomorphizing anecdotes about uxorious ostriches, passionate pike, and prudent alligators. Such excesses inspired a backlash. In 1894, the British psychologist C. Lloyd Morgan established what came to be known as Morgan's canon: "In no case may we interpret an action as the outcome of the exercise of a higher psychical faculty, if it can be interpreted as the outcome of the exercise of one which stands lower in the psychological scale." If instinct could explain why your dog growled at your suitcase, then there was no need to cast about for a richer interpretation, one that might, as Morgan put it, "savour of the prattle of the parlour tea table rather than the sober discussion of the study." As sensible as Morgan's canon sounded, it essentially censored the question "Do animals think?"

The reigning theory for the new era of animal studies—especially in America—was the behaviorism of John Watson and B. F. Skinner. It was firmly rooted in the laboratory, in psychology rather than in biology departments, in the metaphor of animal as machine, and in the use of mazes and the Skinner box. Behaviorists tested the capacities of animals not through naturalistic observation but through highly controlled stimulus response experiments. Speculation about the subjective experiences or thought processes of animals seemed unscientific: animals didn't think, they reacted. As Gregory Radick writes in *The Simian Tongue: The Long Debate About Animal Language* (2007), the idea of animal cognition "came to be regarded as belonging to the sentimental nineteenth century."

Meanwhile, the tale of the horse Clever Hans gave a chill to any scientist still hoping to demonstrate that animals were thinking creatures. Around the turn of the century, Wilhelm von Osten, a German schoolmaster, bought a horse, named it Hans, and supposedly taught him arithmetic—addition, subtraction, even fractions and decimals—along with some spoken and written German. When von Osten asked Hans, for example, "What is twelve divided by three?," Hans tapped his hoof four times. Von Osten gave regular demonstrations of his horse's astonishing abilities, until a psychologist named Oskar Pfungst pointed out that von Osten was unconsciously cuing the animal with subtle movements of his head and eyes. One lesson to have drawn from this episode was that, although Hans may not have been so clever at arithmetic, he was, in fact, quite clever at reading the body language of humans, a talent that could have warranted further investigation. Another—and this was the one that took—was that anybody studying animal communication had to be extremely careful about accidental cuing.

By the mid-nineteen-sixties, the behaviorist paradigm was being challenged by the field of ethology, which emphasized the role of instincts and of underlying biology, and relied more on naturalistic observation of animals than on laboratory experiments. A prominent Harvard zoologist named Donald Griffin—he had discovered that bats navigate through echolocation—began pushing his colleagues to acknowledge that animals had thoughts and emotions. He wrote, "The customary view of animals as always living in a state comparable to that of human sleepwalkers is a sort of negative dogmatism." Wasn't it possible that a chimpanzee who scoured the rain forest for a chunk of granite, then used it to crack open a nut, was consciously thinking about that tasty morsel inside, rather than executing rote movements? And did we really know enough to say whether animals had consciousness or not?

In the late seventies, the ethologists Robert Seyfarth and Dorothy Cheney reported some of the first evidence of what appeared to be a referential communication system among animals. Cheney and Seyfarth had gone to Kenya to study the vervet monkey, which is hunted by multiple predators—pythons, leopards, and eagles. The vervets, it seemed, had developed alarm calls that referred to specific predators—a phenomenon that Cheney and Seyfarth were able to demonstrate by playing recordings of the calls, and watching what the monkeys did upon hearing them. If the call was the one for "Python!," they stood on their hind legs and started looking for it; if the call was

"Leopard!," the vervets climbed the nearest tree; if they were already in a tree and they heard "Eagle!," they scrambled for the bushes.

For some researchers, though, the vital question was not so much "Can animals talk to each other?" as "Can they talk to us?" In the fifties, a chimp named Viki, who had been raised by a psychologist and his wife in their Florida home, and treated as much like a human child as possible, was taught to say four words—"mama," "papa," "cup," and "up"—in a hoarse whisper. And, in the late sixties, Beatrice and Allen Gardner, researchers at the University of Nevada, decided to teach a chimp American Sign Language. Chimps, they reasoned, were notably expressive with their hands, and so a gestural language would probably be easier for them to adopt than a spoken one. The Gardners' success with a chimp named Washoe—she reportedly learned nearly three hundred signs before her death, last year—ushered in a busy era of ape-language studies. In the seventies and eighties, Koko the gorilla, Kanzi the bonobo, and the mischievously named Nim Chimpsky all became simian celebrities. (Noam Chomsky, the renowned linguist, has argued that language is the exclusive endowment of *Homo sapiens.*) The ape studies seemed to appeal to a vague hope that animals might be gurus of sorts, offering humankind a salutary humbling. Francine Patterson, a California-based researcher who worked with Koko, published popular children's books about the gorilla and her love of felines ("soft good cat cat," Koko once signed), and Nim Chimpsky made appearances on "Sesame Street" and on David Susskind's talk show.

The prominent ape studies tended to be time-consuming and strife-ridden—Nim was quite bitter, with the attention span of, well, a chimpanzee—and the results were mixed. Several apes acquired respectable vocabularies, but, unlike most toddlers, they did not learn to produce sentences combining more than two or three words. (A few did seem to comprehend more complicated sentences: Kanzi, who communicates with the aid of a keyboard, reportedly can distinguish, for instance, between the commands "Make the doggie bite the snake" and "Make the snake bite the doggie"—don't worry, these were toys.) When the apes did combine words, the second word was often a nongrammatical intensifier, as in "open hurry." And, when they produced a longer utterance, it tended to be a string of repetitions of the sort rarely encountered outside a Gertrude Stein poem. (A quote from Nim Chimpsky: "Give orange give me eat orange me eat orange give me eat orange give me you.")

In 1979, Herbert Terrace, a psychologist at Columbia University who initiated the Nim Chimpsky research, published an article in *Science* in which he essentially declared the project a failure. Nim knew individual words, and, like Hans the horse, he responded ably to his teachers' cues, but he wasn't really using language. In 1980, a conference in New York seemed to seal Terrace's verdict. Called "The Clever Hans Phenomenon: Communication with Horses, Whales, Apes, and People," it was convened by the semiotician Thomas Sebeok, who invited a magician, the Amazing Randi, and an expert on the psychology of circus animals to make the point that animal-language studies involved an element of deception. Diana Reiss, the dolphin researcher, attended the conference and first met Pepperberg there. Reiss remembers it as a strange and disheartening gathering, in which "Sebeok started off by saying that asking whether animals had language was like asking if elephants could fly."

In the aftermath of the conference, an aura of failure and even chicanery clung to animal-language studies. This was unfortunate, Stephen Anderson suggests in his 2004 book, *Dr. Dolittle's Delusion: Animals and the Uniqueness of Human Language.* The ape-language projects, however flawed, had demonstrated abilities that "had not been previously suspected and about which it would be exciting to learn more." Did apes acquire full-blown expressive language? No. Could they learn to communicate their wants and needs? Yes. Poignantly, even after Nim had been retired to a Texas ranch where most of the employees didn't know sign language, he continued to sign. According to a new book, *Nim Chimpsky: The Chimp Who Would Be Human*, by Elizabeth Hess, when Bob Ingersoll, an old teacher of Nim's, came to visit him, the chimp, who was in a cage at the time, eagerly signed "Bob," "out," and "key." And, when Nim was joined at the ranch by another chimp, Sally, with whom he became quite close, he taught her signs for a couple of his favorite things, such as "gum" and "banana," and signed "sorry" after a quarrel. As Reiss told me, "Nim Chimpsky learned a lot—just not, perhaps, language."

WHEN IRENE PEPPERBERG WENT TO New York for the Clever Hans conference, she was thirty-one, and had owned Alex for three years. She had arrived in the world of animal communication from "out of left field," as Diana Reiss puts it. Pepperberg has a Ph.D. from Harvard in theoretical chemistry, not psychology or zoology. But in the midst of her thesis work, which involved modelling chemical-reaction rates, it suddenly hit her, she recalls, that "(a) we don't know enough at this point to do this exactly right and (b) in the future, what it's taking me seven years to do with a mathematical model is going to take a computer hours, or seconds." She decided to pursue something different. In any case, the prospects for women in her field hadn't been encouraging. Speaking of her class at Harvard, she recalled, "My year was the first year that graduate-school draft deferrals for men were cut way back. So they let in a lot of women for a change. But the women were asked in their job interviews things like 'What kind of birth control are you using?' "

In 1974, Pepperberg watched two programs on PBS, one about Washoe and one about dolphin and whale intelligence. "And it sort of clicked for me that this was real science, that you could be a scientist and study this," she said. "I didn't want to give up science—I loved the scientific method." She also realized that "nobody was doing this kind of research with birds." Pepperberg suspected that this was a mistake, because, unlike apes, some birds were proficient at mimicking human speech. And, based on personal experience, she felt certain that birds could be smart.

Pepperberg was born in 1949, and grew up in an apartment above a store in Brooklyn. She was an only child. Her father, Robert Platzblatt, was a frustrated biochemist who passed on his love of science, bringing home *The Microbe Hunters* and biographies of Marie Curie for his daughter to read. But, when Irene was little, her father was simultaneously teaching middle school in Bedford-Stuyvesant, studying for his master's degree, and taking care of his sick mother. "He'd wake me up and kiss me good morning, and then sometimes I wouldn't see him till the next morning," she recalls. Her mother had been happy working as a junior bookkeeper before she had Irene, and resented staying home, as young mothers were expected to do in the fifties. "For me, there were no other children at home to play with, and for her there was no respite," Pepperberg says.

When Pepperberg was four years old, her father bought her a budgie, to keep her company. She ended up owning a series of budgies, and training them all to speak, at least a little; when she went off to M.I.T. as an undergraduate, one of the birds went with her to her dorm room.

Despite her graduate-school epiphany at Harvard, she continued with her Ph.D. in theoretical chemistry, receiving her degree in 1976; but she also started attending courses in departments relevant to the bird research she now hoped to do. "I was spending forty hours a week learning psychology and biology and forty hours finishing my doctorate," she recalls. In 1977, her husband at the time, David Pepperberg, a neuroscientist, got a professorship at Purdue University, and they moved to Indiana. Pepperberg started looking around for a parrot—preferably an African gray. Among parrot owners, grays have a reputation for being good talkers, and Pepperberg had come across German studies that found them to be adept at numerical and other cognitive tasks. Meanwhile, some recent theorizing about animal intelligence had suggested that the important factor was not absolute brain size but brain size relative to body weight—and on that score, the so-called "encephalization quotient," "birds were surprisingly high up there," Pepperberg said. In her 1999 book, *The Alex Studies*, published by Harvard, Pepperberg writes about becoming excited by an idea that Donald Griffin, the zoologist, had articulated; namely, that "researchers might benefit from studying animals the way that anthropologists study a previously undiscovered primitive tribe." He recommended that scholars attempt "to establish two-way communication and use this communication to determine how they process information and interpret the events in their world."

In 1977, Pepperberg went to a Chicago-area pet store and bought a thirteen-month-old African gray parrot for six hundred dollars. Like all African grays of the Congo subspecies, he was the color of a storm cloud and about a foot tall, with a shiny, scimitar-shaped black beak, a white face, and a lipstick-red tail. She named him Alex. (Luckily for science, when she and her husband divorced, twenty years later, David Pepperberg suggested that she keep the parrot and he keep the dog.)

Pepperberg wasn't sure how to train Alex, although she knew that she didn't want to pursue a behaviorist model. These paradigms had been tried on talking birds, with poor results. (In one 1967 experiment, mynah birds were placed in soundproof boxes where they heard tape-recorded words, followed by the dispensing of food pellets; the birds did not learn to mimic the words. Meanwhile, mynahs that had been left out of the experiment, and turned into pets by the lab assistants, talked fluently.) To Pepperberg, the Skinnerian approach made no sense: "I mean, you don't take a preschooler and put him in a Skinner box! You give him all this enrichment and social interaction." In some ways, a smart bird wasn't so different. Gray parrots are social animals. In the wild, they settle down for the night in roosts with hundreds of other parrots, forage for food in small groups, may have a dominance hierarchy, and mate for life. For mates, imitating one another is a big part of their vocalizing. With their mates, they sing—or squawk—a duet that is unique to that couple. "A single parrot in the wild is a dead parrot," Pepperberg said. "It cannot forage and look for predators at the same time. I was in Australia a number of years ago, and there was this little juvenile rosella at the top of the trees, screaming its head off. Your first reaction is, It's screaming its head off, it's gonna be eaten. And your second reaction is, It might be eaten, but it's got to find its flock."

Pepperberg was further influenced by two papers, one by the primatologist Alison Jolly, in 1966, and one by the psychologist Nicholas Humphrey, in 1976. In different ways, both made the novel argument that the evolutionary pressures fostering intelligence in animals were not the primal demands of subsistence but the more abstract ones of living within a complex social hierarchy. The more intricate the social system, the more evolution would select for intelligence. Pepperberg noted that longevity also played a role. The longer a bird lived, the more it would have to remember—not only about which birds in the flock could dominate which other birds but also about "which trees are blossoming and fruiting, and which trees have died and which paths have now been taken over by leopards, and where the elephants are now going to make their wallows, so it can go and get water."

Pepperberg needed a method for teaching a parrot that played to its particular strengths. She came across something called the model/rival technique, which a German ethologist named Dietmar Todt had tried in a 1975 study of parrots. Todt had reasoned that, since parrots learn to squawk by watching each other vocalize, they might be able to learn German by observing people talk. So he developed a system in which one person was the trainer and one was the model for the bird—and its rival for the trainer's attention. Pepperberg tweaked the protocol: in her version, the model/rival and the trainer periodically exchanged roles, so the bird could see that one person wasn't always in charge. Parrots started the process by learning referential labels for things they wanted, rather than dialogues of the "Hello, how are you? I am fine" variety, which, Pepperberg figured, didn't mean much to a parrot. There were no extrinsic rewards. If the parrot named an object, he'd get to play with that object, and, if he didn't want it, he got the right to ask for something else. Pepperberg explained, "Let's say you're the model/rival and I'm the trainer. I have this object that the bird wants, and I show it to you and I say"—she adopted a singsong voice—"'What's this?,' and you say, 'Cork.' I say, 'That's right,' and you say, 'Cork, cork, cork,' while you're holding it and the bird is practically falling off the perch because he wants it. And he hears that this weird noise is what mediated the transfer of this object. So we change roles, and then, instead of saying 'Cork,' I go, *'Raaaawkk,'*"—an uncannily accurate screech—"and you go, 'No, no, you're wrong,' so the bird sees that not just any weird noise transfers the cork."

The system worked. At first, a parrot might make a sound more like "erk" than "cork." He'd need practice. Certain sounds are nearly impossible to produce without lips—Alex was never able to say "purple," for instance, even after he nailed all his other colors. Still, as Reiss says, "Irene really found the appropriate method based on what we know about these birds. If you can tap into what these birds do in their own environment—in this case, the way these birds pair-bond—then you can set up a powerful learning paradigm."

As it happens, the model/rival method may have some utility for another species: humans. Diane Sherman, who works with autistic children in Monterey, California, has had some preliminary success in encouraging speech in her clients using Pepperberg's protocol. In an article published in *The International Journal of Comparative Psychology*, Sherman and Pepperberg say that, in two studies of children in Sherman's private practice, the model/rival method led to "significant gains" in the children's "communication and social interaction with peers and adults." (Behavioral changes were measured by reports from parents and teachers, and included criteria like demonstrations of empathy, improved eye contact, saying hello to people, and speaking in sentences.)

When Pepperberg started publishing papers on Alex, Reiss recalls, many of her colleagues were nonplussed: "I can remember being at a psychology conference in the early eighties where Irene was giving a paper about Alex, and people were saying, 'I see it but I don't believe it.' A lot of the work being presented was pretty Skinnerian and I remember that, at her talk, some people in the back row got up and walked out. Now she gets a radically different reception. I was at a comparative-cognition conference recently, and many of those same people were there. And I heard a number of people talking about Irene's work and referring to it as groundbreaking."

A FEW WEEKS AFTER PEPPERBERG'S speech at the Bird Expo, I saw her in a more properly scientific guise, teaching an undergraduate seminar and an evening extension course on animal cognition at Harvard. At the evening course, she lectured about one of her favorite subjects—the songbirds known as oscines, which learn their melodies much like human children learn language. They have an innate predisposition to learn song; they have an initial period of babbling; they have specific brain areas for song-learning, just as we have for speech; and they have a sensitive phase in childhood for acquiring song, just as humans have for acquiring language. The parallel is striking, Pepperberg told her students, partly because so few other species actually learn to communicate vocally, as opposed to making primal sounds instinctually. "Besides humans, it's dolphins and certain birds—and maybe, we think now, elephants," she said.

An innate predisposition isn't enough for oscine songbirds to learn their repertoire, Pepperberg said. They need tutoring from members of their own species—usually, from their fathers or the dominant males in their flock, since it is males who tend to sing, in order to attract mates or defend territory. Pepperberg spoke of songbirds with prodigious repertoires—the adult brown thrasher can sing two thousand songs. In a voice that had a faint trace of her native Brooklyn, she said, "You can raise some birds in social isolation and give them audiotapes of their own species singing, but it doesn't work very well. And, if you keep them alone in a little tiny box during their sensitive phase, they will sing, but it will be a kind of tentative, crummy song." She urged her students to "go out and listen to the dawn chorus. It's cacophony. A beautiful cacophony, but nonetheless." She argued that, in order for a songbird to recognize the music of its own species, it must be able to make distinctions between same and different—just as Alex did.

In speaking about animal intelligence, Pepperberg has tried to strike a balance between what the ecologist James Gould has called "the unprofitable extremes of blinding skepticism and crippling romanticism." Pepperberg has published widely in peer-reviewed scientific journals, even as she raises funds for her research with a Web site that sells adorable Alex tote bags, key chains, and mugs.

Still, for many years, Pepperberg felt that she and Alex were, in intellectual terms, out on a limb. In the past decade, though, dozens of studies have buttressed Pepperberg's claims about avian intelligence. Alex's brain was the size of a shelled walnut, as Pepperberg often observed. Yet Nathan Emery, a cognitive biologist at the University of London, points out that, "in relation to their body size, parrots have brains as big as those of chimpanzees." Emery has taken to calling certain of the brighter birds the "feathered apes in your garden." And Erich Jarvis, a neuroscientist at Duke University, argues that avian brains, long regarded as primitive, are not so different from mammalian brains after all. Birds, Jarvis explains, "have a cortical region that developed, in fact, from the same substrate as in humans." Soon enough, "birdbrain" may no longer be a viable insult.

One important set of studies centers on the clever corvid family of birds, which includes crows, ravens, jays, and magpies. (Perhaps it's no coincidence that crows or ravens often appear as cunning tricksters and problem-solvers in Native American legends and Aesop's fables.) A 1998 study suggested that these birds have a surprising capacity for "episodic-like memory." Scrub jays hide food and retrieve it later, and in the study the birds were allowed to cache two kinds of food: peanuts and wax-moth larvae, which they much prefer (though only when the larvae are fresh). Some of the jays were then sent to their cages for four hours, and some for five days, after which they were given access to their stashes. The birds that were released first went for the larvae, while the ones let out later settled for the peanuts, apparently having remembered not only when and where they had hidden their food but how long it took for the larvae to go bad. Further studies noted that, if a scrub jay realized that it was being spied on by another jay as it cached its food, it was more likely to come back later and hide the goodies elsewhere. Moreover, jays that had previously stolen from another's stash were more prone to move their food to a different hiding place. (It takes a thief to know one, apparently.) Such behavior suggests that jays not only have a Machiavellian streak; they also possess a "theory of mind," and can guess, to some degree, what other birds are thinking. The husband-and-wife team of Nicola Clayton and Nathan Emery, who are largely responsible for this work, argue that it shows some birds have "the same cognitive tool kit" that apes have: "causal reasoning, flexibility, imagination, and prospection." They believe that evolution created multiple paths to intelligence—and that "complex cognitive abilities had evolved multiple times in distantly related species."

Alex Kacelnik, a professor at Oxford, studies another corvid, the New Caledonian crow, which has a rare ability to make tools (a talent once thought to be limited to primates). It will tear a strip off a leaf, and then use the strip to poke into insect-harboring crevices. In 2002, Kacelnik and his colleagues presented a group of birds with the challenge of making a tool, with a material not found in nature, to solve a novel problem. In one room, which the birds could enter freely, was a length of wire and a cylindrical container that had something tasty, like a piece of pig's heart, at the bottom. One of the crows, a female named Betty, figured out how to bend the wire with her beak, so as to fashion a hook for retrieving the meat. Did Betty look at the wire and have an insight about what to do? Not exactly. "The crows don't fully understand what they are doing," Kacelnik told me. "And yet neither can you say it's explainable as purely mechanical repetition." To insure that Betty wasn't somehow acting on instinct, Kacelnik and his colleagues "gave her a task that was the opposite. We gave her a piece of metal that was bent so it was too short. Would she be clever enough to unbend it? She was. So that implied some understanding of the physical requirements of the task." Though some crows, like Betty, cracked the challenge quickly, others took many tries; still others never mastered it. Watching videos of Betty on Kacelnik's Web site, I noticed that she seemed to have a particularly focussed and alert way about her. Even Kacelnik, who is loath to anthropomorphize, confessed to me, "An element of our finding that still puzzles me is that while Betty was not chosen or treated in any special way, she was different. She showed a readiness to coöperate and solve problems that none of the other animals in our study have replicated. We have no idea why."

How representative are Betty and Alex of their species? For science, one remarkable bird is not enough. Kacelnik told me, "We study multiple animals for a reason: if you want to characterize a species, you need to try similar ideas on different subjects, to examine how much is due to exceptional circumstances and how much is a property of your method and your experiments." I asked him why more researchers weren't working with African grays, trying to replicate Pepperberg's achievements with Alex. "The problem with these animals is that they are the opposite of fruit flies," he said, meaning that parrots live a long time—often, fifty to sixty years in captivity. "Alex was still learning when he died, and he was thirty." He later elaborated: "Irene's work could not really have been planned ahead, as nobody knew what was possible. Alex's development as a unique animal accompanied Irene's as a unique scientist. Hers is not a career trajectory one would advise to young scientists—it's too risky."

Pepperberg told me that there are now a few groups of researchers studying cognition in African grays: one group in Prague, where the parrots are learning Czech; one in Milan, where they are learning Italian; and one outside Paris, where they are studying "referential communication" that does not involve language. In the end, though, it may be Pepperberg herself who has the greatest drive and capacity to replicate her own work, although she has suffered from funding shortfalls; over the decades, she has received eight grants from the National Science Foundation, but she does a lot of fund-raising on her own. At the Brandeis lab, she is training and testing her two remaining parrots, Griffin and Arthur. Griffin, who is thirteen, shows more promise than Arthur, who is nine. Both seem shyer and less confident than Alex. Pepperberg's lab manager, Arlene Levin, describes Griffin as "a timid guy, and a plucker." (Anxious parrots pluck out their feathers.) And Griffin grew up in Alex's shadow; he was the one whom Alex interrupted and corrected—"Say better," "Pay attention," "Bad parrot." Griffin currently knows about a dozen words for objects; Alex knew fifty. "I'm not saying that Griffin won't get there," Pepperberg told me. "I think he will. He's smart. It's just a matter of convincing him that, no, Alex is not going to give you the answer. Don't just sit there waiting!" She admits that there's pressure on her, too: "It's important to show that Alex was not some unique individual."

Even Nathan Emery, the cognitive biologist, who is otherwise admiring of Pepperberg's work, thinks it is possible that Alex might have been an exceptionally bright bird. Pepperberg notes that she picked him at random at the pet store, precisely to forestall the suspicion that she'd gone hunting for the avian Einstein. She says, "I think that what made Alex special was that, for the first fifteen years of his life, he was an only bird, and he had people with him talking with him eight hours a day—interacting with him, treating him like a toddler. When he said something, you reacted in some way or another, so he learned that his vocalizations could control the environment."

Clive Wynne, an animal psychologist at the University of Florida, is suspicious of most animal-communication projects. (He writes papers with titles like "The Perils of Anthropomorphism" and "Pets Aren't Us.") He is no less dubious about Pepperberg's work. "An African gray parrot is an expensive pet, but it's not expensive compared with, say, a dolphin, which costs tens of thousands of dollars," he says. "Why is Alex the only parrot in the history of parrotdom to have done these things? If there's really something going on here—I mean, he lived thirty years—then somebody else, somewhere along the line, would have replicated this. If we really want this to be science and not just some sort of adjunct to the entertainment industry, we shouldn't be relying on one animal."

LAST MONTH, I VISITED PEPPERBERG and her two remaining birds at Brandeis. Her lab is a windowless, fluorescent-lit room, ten feet by fifteen feet, with white cement walls. One wall is lined with shelves stacked with bags of nuts and kibbles, boxes of bite-size Shredded Wheat, and plastic jars full of colored wooden blocks, plastic letters, and pompoms. There are newspapers on the floor, and three big cages, one of them empty. Alex, it's clear, was not simply a pet or a laboratory subject. For thirty years, he was the center of Pepperberg's life: a little like a child, a little like a significant other, very much like a collaborator. When Alex died—a technician discovered his body—she was devastated. "I always felt I was working toward the next stage with him," she told me. "And I thought I had at least a decade more." When two close friends of Pepperberg's—a nuclear-submarine commander named Carl Hartsfield and his wife, Leigh Ann—heard the news, they drove up from Washington, D.C., along with their own African gray, Pepper, to be with her.

My first glimpse of Griffin made me glad that I'm not a parrot who has to identify my life's mate in a flock. Griffin, who was shuffling down one side of his cage in a parroty version of the Electric Slide, looked just like the photographs of Alex that I'd seen, and just like Arthur, who was over in a corner, quietly preening. Griffin is named, in part, for Donald Griffin, and in part for the gryphon he reminded Pepperberg of when she first bought him—"He was all talons and beak." Now he's grown into himself, and, when he's not plucking, is full-feathered and handsome, just like Alex. I did notice that, when Griffin spoke, his voice was softer than the Great One's, which is preserved on dozens of YouTube clips.

One of the odd things about observing Griffin speak was the contrast between his face—goggle-eyed and masklike, and much less expressive than a dog's—and his ability to make his wishes clear. "Wanna go chair," Griffin said, and Arlene Levin, the lab manager, picked him up and placed him on a perch next to her office chair. Then Levin showed me a little of what the new boy could do. "What matter?" she asked, holding up a small block of wood close to Griffin's beak, so he could tap it if he wanted. This was how the parrots were queried about what material an object is made of.

"Wood," Griffin said.

"You want the wood?"

Griffin didn't pick up the block.

"What do you want?"

"Nut." Griffin was given an almond, and he snarfed it down.

"What matter?" Levin asked, holding a stone.

"Rrrr-ock," Griffin said.

"Want the rock?"

No—he wanted another nut. (Nuts are offered only in training sessions.)

"What color?" Levin asked, showing him a plastic cup.

"Green," Griffin announced correctly, in an uninflected tone, and then "yellow" when shown another cup.

A while later, two Brandeis undergraduates, Shannon Cabell and Michelle Barras, came in to do a training session, and it didn't go quite as smoothly. Pepperberg was sitting at a desk in the corner, thumbing through the latest issue of *Animal Cognition*. The two young women were perched on tall stools, and Griffin was on a perch across from them.

"What shape?" Cabell asked. She had to repeat it four or five times, as Griffin wasn't saying a thing.

From the corner, Pepperberg instructed, "All right, model it for him." Cabell picked up a square object and asked Barras what shape she was holding. With a theatrical display of pride, Barras responded, "Four-corner!"

"Right," Cabell said. "She's such a good bird. She gets a four-corner for that."

"Wanna nut," Barras said, in convincing parrotese. Cupping the square in her hand, she gamely mimed eating it.

"What color?" Cabell asked about a yellow pompom, a material that the birds identify as wool.

"Wool," Griffin chimed.

Pepperberg sighed, distractedly pulled a piece of sandwich bread out of a plastic bag on the shelf, and began eating it. "Right," she told Griffin. "It's wool. What *color* wool?"

"Yellow." There were a few more exchanges like this, as well as some long silences during which Pepperberg murmured, "C'mon, Griffs," and Barras said, "All these nuts can be yours."

Watching this exchange, I indulged in my own anthropomorphic speculation—I imagined that Griffin was thinking that he'd like to be back where he was half an hour ago, sitting quietly in Pepperberg's lap, having his head tickled.

Barras pulled out a green block and asked Griffin what color it was. The parrot stared at his feet.

"See if he'll do it for a bean," Pepperberg suggested. Barras took an orange jelly bean out of a bag. "No," Pepperberg prompted. "A bean the same color." Barras dug out a green one. In the fall, candy corn is the major training treat; in the winter, it's candy hearts; in the spring, there are jelly beans; in the summer, the parrots get Skittles. "Half a jelly bean once in a while is acceptable for his diet," Pepperberg told me. It's also a powerful motivator.

"*Green,*" Griffin said immediately.

This parrot is good, even impressive. But he's not Alex. Not yet.

GARY WOLF

Want to Remember Everything You'll Ever Learn? Surrender to This Algorithm

FROM *WIRED*

Want to improve your memory significantly? A reclusive genius has figured out a simple way to do it. As Gary Wolf discovers, the method may be simple; actually using it, however, isn't easy.

THE WINTER SUN SETS IN MID-AFTERNOON IN KOLOBRZEG, Poland, but the early twilight does not deter people from taking their regular outdoor promenade. Bundled up in parkas with fur-trimmed hoods, strolling hand in mittened hand along the edge of the Baltic Sea, off-season tourists from Germany stop openmouthed when they see a tall, well-built, nearly naked man running up and down the sand.

"Kalt? Kalt?" one of them calls out. The man gives a polite but vague answer, then turns and dives into the waves. After swimming back and forth in the 40-degree water for a few minutes, he emerges from the surf and jogs briefly along the shore. The wind is strong, but the man makes no move to get dressed. Passersby continue to comment and stare. "This is one of the reasons I prefer anonymity," he tells me in English. "You do something even slightly out of the ordinary and it causes a sensation."

Piotr Wozniak's quest for anonymity has been successful. Nobody along this string of little beach resorts recognizes him as the inventor of a technique to turn people into geniuses. A portion of this technique, embodied in a software program called SuperMemo, has enthusiastic users around the world. They apply it mainly to learning languages, and it's popular among people for whom fluency is a necessity—students from Poland or other poor countries aiming to score well enough on English-language exams to study abroad. A substantial number of them do not pay for it, and pirated copies are ubiquitous on software bulletin boards in China, where it competes with knockoffs like SugarMemo.

SuperMemo is based on the insight that there is an ideal moment to practice what you've learned. Practice too soon and you waste your time. Practice too late and you've forgotten the material and have to relearn it. The right time to practice is just at the moment you're about to forget. Unfortunately, this moment is different for every person and each bit of information. Imagine a pile of thousands of flash cards. Somewhere in this pile are the ones you should be practicing right now. Which are they?

Fortunately, human forgetting follows a pattern. We forget exponentially. A graph of our likelihood of getting the correct answer on a quiz sweeps quickly downward over time and then levels off. This pattern has long been known to cognitive psychology, but it has been difficult to put to practical use. It's too complex for us to employ with our naked brains.

Twenty years ago, Wozniak realized that computers could easily calculate the moment of forgetting if he could discover the right algorithm. SuperMemo is the result of his research. It predicts the future state of a person's memory and schedules information reviews at the optimal time. The effect is striking. Users can seal huge quantities of vocabulary into their brains. But for Wozniak, 46, helping people learn a foreign language fast is just the tiniest part of his goal. As we plan the days, weeks, even years of our lives, he would have us rely not merely on our traditional sources of self-knowledge—introspection, intuition, and conscious thought—but also on something new: predictions about ourselves encoded in machines.

Given the chance to observe our behaviors, computers can run simulations, modeling different versions of our path through the world. By tuning these models for top performance, computers will give us rules to live by. They will be able to tell us when to wake, sleep, learn, and exercise; they will cue us to remember what we've read, help us track whom we've met, and remind us of our goals. Computers, in Wozniak's scheme, will increase our intellectual capacity and enhance our rational self-control.

The reason the inventor of SuperMemo pursues extreme anonymity, asking me to conceal his exact location and shunning even casual recognition by users of his software, is not because he's paranoid or a misanthrope but because he wants to avoid random interruptions to a long-running experiment he's conducting on himself. Wozniak is a kind of algorithmic man. He's exploring what it's like to live in strict obedience to reason. On first encounter, he appears to be one of the happiest people I've ever met.

IN THE LATE 1800S, A German scientist named Hermann Ebbinghaus made up lists of nonsense syllables and measured how long it took to forget and then relearn them. (Here is an example of the type of list he used: bes dek fel gup huf jeik mek meun pon daus dor gim ke4k be4p bCn hes.) In experiments of breathtaking rigor and tedium, Ebbinghaus practiced and recited from memory 2.5 nonsense syllables a second, then rested for a bit and started again. Maintaining a pace of rote mental athleticism that all students of foreign verb conjugation will regard with awe, Ebbinghaus trained this way for more than a year. Then, to show that the results he was getting weren't an accident, he repeated the entire set of experiments three years later. Finally, in 1885, he published a monograph called *Memory: A Contribution to Experimental Psychology.* The book became the founding classic of a new discipline.

Ebbinghaus discovered many lawlike regularities of mental life. He was the first to draw a learning curve. Among his original observations was an account of a strange phenomenon that would drive his successors half batty for the next century: the spacing effect.

Ebbinghaus showed that it's possible to dramatically improve learning by correctly spacing practice sessions. On one level, this finding is trivial; all students have been warned not to cram. But the efficiencies created by precise spacing are so large, and the improvement in performance so predictable, that from nearly the moment Ebbinghaus described the spacing effect, psychologists have been urging educators to use it to accelerate human progress. After all, there is a tremendous amount of material we might want to know. Time is short.

However, this technique never caught on. The spacing effect is "one of the most remarkable phenomena to emerge from laboratory research on learning," the psychologist Frank Dempster wrote in 1988, at the beginning of a typically sad encomium published in *American Psychologist* under the title "The Spacing Effect: A Case Study in the Failure to Apply the Results of Psychological Research." The sorrowful tone is not hard to understand. How would computer scientists feel if people continued to use slide rules for engineering calculations? What if, centuries after the invention of spectacles, people still dealt with nearsightedness by holding things closer to their eyes? Psychologists who studied the spacing effect thought they possessed a solution to a problem that had frustrated humankind since before written language: how to remember what's been learned. But instead, the spacing effect became a reminder of the impotence of laboratory psychology.

As a student at the Poznan University of Technology in western Poland in the 1980s, Wozniak was overwhelmed by the sheer number of things he was expected to learn. But that wasn't his most troubling problem. He wasn't just trying to pass his exams; he was trying to learn. He couldn't help noticing that within a few months of completing a class, only a fraction of the knowledge he had so painfully acquired remained in his mind. Wozniak knew nothing of the spacing effect, but he knew that the methods at hand didn't work.

The most important challenge was English. Wozniak refused to be satisfied with the broken, half-learned English that so many otherwise smart students were stuck with. So he created an analog database, with each entry consisting of a question and answer on a piece of paper. Every time he reviewed a word, phrase, or fact, he meticulously noted the date and marked whether he had forgotten it. At the end of the session, he tallied the number of remembered and forgotten items. By 1984, a century after Ebbinghaus finished his second series of experiments on nonsense syllables, Wozniak's database contained 3,000 English words and phrases and 1,400 facts culled from biology, each with a complete repetition history. He was now prepared to ask himself an important question: How long would it take him to master the things he wanted to know?

The answer: too long. In fact, the answer was worse than too long. According to Wozniak's first calculations, success was impossible. The problem wasn't learning the material; it was retaining it. He found that 40 percent of his English vocabulary vanished over time. Sixty percent of his biology answers evaporated. Using some simple calculations, he figured out that with his normal method of study, it would require two hours of practice every day to learn and retain a modest English vocabulary of 15,000 words. For 30,000 words, Wozniak would need twice that time. This was impractical.

Wozniak's discouraging numbers were roughly consistent with the results that Ebbinghaus had recorded in his own experiments and that have been confirmed by other psychologists in the decades since. If students nonetheless manage to become expert in a few of the things they study, it's not because they retain the material from their lessons but because they specialize in a relatively narrow subfield where intense practice keeps their memory fresh. When it comes to language, the received wisdom is that immersion—usually amounting to actual immigration—is necessary to achieve fluency. On one hand, this is helpful advice. On the other hand, it's an awful commentary on the value of countless classroom hours. Learning things is easy. But remembering them—this is where a certain hopelessness sets in.

As Wozniak later wrote in describing the failure of his early learning system: "The process of increasing the size of my databases gradually progressed at the cost of knowledge retention." In other words, as his list grew, so did his forgetting. He was climbing a mountain of loose gravel and making less and less progress at each step.

THE PROBLEM OF FORGETTING MIGHT not torment us so much if we could only convince ourselves that remembering isn't important. Perhaps the things we learn—words, dates, formulas, historical and biographical details—don't really matter. Facts can be looked up. That's what the Internet is for. When it comes to learning, what really matters is how things fit together. We master the stories, the schemas, the frameworks, the paradigms; we rehearse the lingo; we swim in the episteme.

The disadvantage of this comforting notion is that it's false. "The people who criticize memorization—how happy would they be to spell out every letter of every word they read?" asks Robert Bjork, chair of UCLA's psychology department and one of the most eminent memory researchers. After all, Bjork notes, children learn to read whole words through intense practice, and every time we enter a new field we become children again. "You can't escape memorization," he says. "There is an initial process of learning the names of things. That's a stage we all go through. It's all the more important to go through it rapidly." The human brain is a marvel of associative processing, but in order to make associations, data must be loaded into memory.

Once we drop the excuse that memorization is pointless, we're left with an interesting mystery. Much of the information does remain in our memory, though we cannot recall it. "To this day," Bjork says, "most people think about forgetting as decay, that memories are like footprints in the sand that gradually fade away. But that has been disproved by a lot of research. The memory appears to be gone because you can't recall it, but we can prove that it's still there. For instance, you can still recognize a 'forgotten' item in a group. Yes, without continued use, things become inaccessible. But they are not gone."

After an ichthyologist named David Starr Jordan became the first president of Stanford University in the 1890s, he bequeathed to memory researchers one of their favorite truisms: Every time he learned the name of a student, Jordan is said to have complained, he forgot the name of a fish. But the fish to which Jordan had devoted his research life were still there, somewhere beneath the surface of consciousness. The difficulty was in catching them.

During the years that Wozniak struggled to master English, Bjork and his collaborator, Elizabeth Bjork (she is also a professor of psychology; the two have been married since 1969), were at work on a new theory of forgetting. Both were steeped in the history of laboratory research on memory, and one of their goals was to get to the bottom of the spacing effect. They were also curious about the paradoxical tendency of older memories to become stronger with the passage of time, while more recent memories faded. Their explanation involved an elegant model with deeply counterintuitive implications.

Long-term memory, the Bjorks said, can be characterized by two components, which they named retrieval strength and storage strength. Retrieval strength measures how likely you are to recall something right now, how close it is to the surface of your mind. Storage strength measures how deeply the memory is rooted. Some memories may have high storage strength but low retrieval strength. Take an old address or phone number. Try to think of it; you may feel that it's gone. But a single reminder could be enough to restore it for months or years. Conversely, some memories have high retrieval strength but low storage strength. Perhaps you've recently been told the names of the children of a new acquaintance. At this moment they may be easily accessible, but they are likely to be utterly forgotten in a few days, and a single repetition a month from now won't do much to strengthen them at all.

The Bjorks were not the first psychologists to make this distinction, but they and a series of collaborators used a broad range of experimental data to show how these laws of memory wreak havoc on students and teachers. One of the problems is that the amount of storage strength you gain from practice is inversely correlated with the current retrieval strength. In other words, the harder you have to work to get the right answer, the more the answer is sealed in memory. Precisely those things that seem to signal we're learning well—easy performance on drills, fluency during a lesson, even the subjective feeling that we know something—are misleading when it comes to predicting whether we will remember it in the future. "The most motivated and innovative teachers, to the extent they take current performance as their guide, are going to do the wrong things," Robert Bjork says. "It's almost sinister."

The most popular learning systems sold today—for instance, foreign language software like Rosetta Stone—cheerfully defy every one of the psychologists' warnings. With its constant feedback and easily accessible clues, Rosetta Stone brilliantly creates a sensation of progress. "Go to Amazon and look at the reviews," says Greg Keim, Rosetta Stone's CTO, when I ask him what evidence he has that people are really remembering what they learn. "That is as objective as you can get in terms of a user's sense of achievement." The sole problem here, from the psychologists' perspective, is that the user's sense of achievement is exactly what we should most distrust.

The battle between lab-tested techniques and conventional pedagogy went on for decades, and it's fair to say that the psychologists lost. All those studies of human memory in the lab—using nonsense syllables, random numbers, pictures, maps, foreign vocabulary, scattered dots—had so little influence on actual practice that eventually their irrelevance provoked a revolt. In the late '70s, Ulric Neisser, the pioneering researcher who coined the term cognitive psychology, launched a broad attack on the approach of Ebbinghaus and his scientific kin.

"We have established firm empirical generalizations, but most of them are so obvious that every 10-year-old knows them anyway," Neisser complained. "We have an intellectually impressive group of theories, but history offers little confidence that they will provide any meaningful insight into natural behavior." Neisser encouraged psychologists to leave their labs and study memory in its natural environment, in the style of ecologists. He didn't doubt that the laboratory theories were correct in their limited way, but he wanted results that had power to change the world.

Many psychologists followed Neisser. But others stuck to their laboratory methods. The spacing effect was one of the proudest lab-derived discoveries, and it was interesting precisely because it was not obvious, even to professional teachers. The same year that Neisser revolted, Robert Bjork, working with Thomas Landauer of Bell Labs, published the results of two experiments involving nearly 700 undergraduate students. Landauer and Bjork were looking for the optimal moment to rehearse something so that it would later be remembered. Their results were impressive: The best time to study something is at the moment you are about to forget it. And yet—as Neisser might have predicted—that insight was useless in the real world. Determining the precise moment of forgetting is essentially impossible in day-to-day life.

Obviously, computers were the answer, and the idea of using them was occasionally suggested, starting in the 1960s. But except for experimental software, nothing was built. The psychologists were interested mainly in theories and models. The teachers were interested in immediate signs of success. The students were cramming to pass their exams. The payoff for genuine progress was somehow too abstract, too delayed, to feed back into the system in a useful way. What was needed was not an academic psychologist but a tinkerer, somebody with a lot of time on his hands, a talent for mathematics, and a strangely literal temperament that made him think he should actually recall the things he learned.

The day I first meet Wozniak, we go for a 7-mile walk down a windy beach. I'm in my business clothes and half comatose from jet lag; he's wearing a track suit and comes toward me with a gait so buoyant he seems about to take to the air. He asks me to walk on the side away from the water. "People say that when I get excited I tend to drift in their direction, so it is better that I stand closer to the sea so I don't push you in," he says.

Wozniak takes an almost physical pleasure in reason. He loves to discuss things with people, to get insight into their personalities, and to give them advice—especially in English. One of his most heartfelt wishes is that the world have one language and one currency so this could all be handled more efficiently. He's appalled that Poland is still not in the Eurozone. He's baffled that Americans do not use the metric system. For two years he kept a diary in Esperanto.

Although Esperanto was the ideal expression of his universalist dreams, English is the leading real-world implementation. Though he has never set foot in an English-speaking country, he speaks the language fluently. "Two words that used to give me trouble are *perspicuous* and *perspicacious*," he confessed as we drank beer with raspberry syrup at a tiny beachside restaurant where we were the only customers. "Then I found a mnemonic to enter in SuperMemo: clear/clever. Now I never misuse them."

Wozniak's command of English is the result of a series of heroic experiments, in the tradition of Ebbinghaus. They involved relentless sessions of careful self-analysis, tracked over years. He began with the basic conundrum of too much to study in too little time. His first solution was based on folk wisdom. "It is a common intuition," Wozniak later wrote, "that with successive repetitions, knowledge should gradually become more durable and require less frequent review."

This insight had already been proven by Landauer and Bjork, but Wozniak was unaware of their theory of forgetting or of any of the landmark studies in laboratory research on memory. This ignorance was probably a blessing, because it forced him to rely on pragmatic engineering. In 1985, he divided his database into three equal sets and created schedules for studying each of them. One of the sets he studied every five days, another every 18 days, and the third at expanding intervals, increasing the period between study sessions each time he got the answers right.

This experiment proved that Wozniak's first hunch was too simple. On none of the tests did his recall show significant improvement over the naive methods of study he normally used. But he was not discouraged and continued making ever more elaborate investigations of study intervals, changing the second interval to two days, then four days, then six days, and so on. Then he changed the third interval, then the fourth, and continued to test and measure, measure and test, for nearly a decade. His conviction that forgetting could be tamed by following rules gave him the intellectual fortitude to continue searching for those rules. He doggedly traced a matrix of paths, like a man pacing off steps in a forest where he is lost.

All of his early work was done on paper. In the computer science department at the Poznan University of Technology, "we had a single mainframe of Polish-Russian design, with punch cards," Wozniak recalls. "If you could stand in line long enough to get your cards punched, you could wait a couple of days more for the machine to run your cards, and then at last you got a printout, which was your output."

The personal computer revolution was already pretty far along in the US by the time Wozniak managed to get his hands on an Amstrad

PC 1512, imported through quasi-legal means from Hamburg, Germany. With this he was able to make another major advance in SuperMemo—computing the difficulty of any fact or study item and adjusting the unique shape of the predicted forgetting curve for every item and user. A friend of Wozniak's adapted his software to run on Atari machines, and as access to personal computers finally spread among students, so did SuperMemo.

After the collapse of Polish communism, Wozniak and some fellow students formed a company, SuperMemo World. By 1995, their program was one of the most successful applications developed by the country's fledgling software industry, and they were searching for funding that would allow them to relocate to Silicon Valley. That year, at Comdex in Las Vegas, 200,000 people got a look at Sony's new DVD technology, prototypes of flatscreens, and Wozniak's SuperMemo, which became the first Polish product shown at the great geek carnival, then at the height of its influence. In Europe, the old communist experiment in human optimization had run its course. Wozniak believed that in a world of open competition, where individuals are rewarded on merit, a scientific tool that accelerated learning would find customers everywhere.

WOZNIAK'S CHIEF PARTNER IN THE campaign to reprogram the world's approach to learning through SuperMemo was Krzysztof Biedalak, who had been his classmate at the University of Technology. The two men used to run six miles to a nearby lake for an icy swim. Biedalak agrees with Wozniak that winter swimming is good for mental health. Biedalak also agrees with Wozniak that SuperMemo produces extreme learning. But Biedalak does not agree with Wozniak about everything. "I don't apply his whole technique," he says. "In my context, his technique is inapplicable."

What Biedalak means by Wozniak's technique is the extension of algorithmic optimization to all dimensions of life. Biedalak is CEO of SuperMemo World, which sells and licenses Wozniak's invention.

Today, SuperMemo World employs just 25 people. The venture capital never came through, and the company never moved to California. About 50,000 copies of SuperMemo were sold in 2006, most for less than $30. Many more are thought to have been pirated.

Biedalak and I meet and talk in a restaurant in downtown Warsaw where the shelves are covered in gingham and the walls are lined with jars of pickled vegetables. He has an intelligent, somewhat hangdog expression, like a young Walter Matthau, and his tone is as measured as Wozniak's is impulsive. Until I let the information slip, he doesn't even know the exact location of his partner and friend.

"Piotr would never go out to promote the product, wouldn't talk to journalists, very rarely agreed to meet with somebody," Biedalak says. "He was the driving force, but at some point I had to accept that you cannot communicate with him in the way you can with other people."

The problem wasn't shyness but the same intolerance for inefficient expenditure of mental resources that led to the invention of SuperMemo in the first place. By the mid-'90s, with SuperMemo growing more and more popular, Wozniak felt that his ability to rationally control his life was slipping away. "There were 80 phone calls per day to handle. There was no time for learning, no time for programming, no time for sleep," he recalls. In 1994, he disappeared for two weeks, leaving no information about where he was. The next year he was gone for 100 days. Each year, he has increased his time away. He doesn't own a phone. He ignores his email for months at a time. And though he holds a PhD and has published in academic journals, he never attends conferences or scientific meetings.

Instead, Wozniak has ridden SuperMemo into uncharted regions of self-experimentation. In 1999, he started making a detailed record of his hours of sleep, and now he's working to correlate that data with his daily performance on study repetitions. Psychologists have long believed there's a correlation between sleep and memory, but no mathematical law has been discovered. Wozniak has also invented a way to apply his learning system to his intake of unstructured information from books and articles, winnowing written material down to the type of discrete chunks that can be memorized, and then scheduling them for efficient learning. He selects a short section of what he's reading and copies it into the SuperMemo application, which predicts when he'll want to read it again so it sticks in his mind. He cuts and pastes completely unread material into the system, assigning it a priority. SuperMemo shuffles all his potential knowledge into a queue and presents it to him on a study screen when the time is right. Wozniak can look at a graph of what he's got lined up to learn and adjust the priority rankings if his goals change.

These techniques are designed to overcome steep learning curves through automated steps, like stairs on a hill. He calls it incremental reading, and it has come to dominate his intellectual life. Wozniak no longer wastes time worrying that he hasn't gotten to some article he wants to read; once it's loaded into the system, he trusts his algorithm to apportion it to his consciousness at the appropriate time.

The appropriate time, that is, for him. Having turned over his mental life to a computerized system, he refuses to be pushed around by random inputs and requests. Naturally, this can be annoying to people whose messages tend to sift to the bottom. "After four months," Biedalak says sadly, "you sometimes get a reply to some sentence in an email that has been scrambled in his incremental reading process."

For Wozniak, these misfires were less a product of scrambling than of an inevitable clash of goals. A person who understands the exact relationship between learning and time is forced to measure out his hours with a certain care. SuperMemo was like a genie that granted Wozniak a wish: unprecedented power to remember. But the value of what he remembered depended crucially on what he studied, and what he studied depended on his goals, and the selection of his goals rested upon the efficient acquisition of knowledge, in a regressive function that propelled him relentlessly along the path he had chosen. The guarantee that he would not forget what he learned was both a gift and a demand, requiring him to sacrifice every extraneous thing.

From the business side of SuperMemo, Wozniak's priorities can sometimes look selfish. Janusz Murakowski, one of Wozniak's friends who worked as a manager at the company during its infancy, thinks that Wozniak's focus on his own learning has stunted the development of his invention. "Piotr writes this software for himself," says Murakowski, now a professor of electrical engineering at the University of Delaware. "The interface is just impossible." This is perhaps a bit unfair. SuperMemo comes in eight flavors, some of which were coded by licensees: SuperMemo for Windows, for Palm devices, for several cell phones, even an Internet version. It's true that Wozniak is no Steve Jobs, and his software has none of the viral friendliness of a casual game like *Brain Age* for Nintendo DS. Still, it can hardly be described as the world's most difficult program. After all, photographers can learn to produce the most arcane effects in Photoshop. Why shouldn't more people be able to master SuperMemo?

"It was never a feel-good product," Murakowski says, and here he may be getting closer to the true conflict that lies at the heart of the struggle to optimize intelligence, a conflict that transcends design and touches on some curious facts about human nature. We are used to the idea that normal humans can perform challenging feats of athleticism. We all know someone who has run a marathon or ridden a bike cross-country. But getting significantly smarter—that seems to be different. We associate intelligence with pure talent, and academic learning with educational experiences dating far back in life. To master a difficult language, to become expert in a technical field, to make a scientific contribution in a new area—these seem like rare things. And so they are, but perhaps not for the reason we assume.

The failure of SuperMemo to transform learning uncannily repeats the earlier failures of cognitive psychology to influence teachers and students. Our capacity to learn is amazingly large. But optimal learning demands a kind of rational control over ourselves that does not come easily. Even the basic demand for regularity can be daunting. If you skip a few days, the spacing effect, with its steady march of sealing knowledge in memory, begins to lose its force.

Progress limps. When it comes to increasing intelligence, our brain is up to the task and our technology is up to the task. The problem lies in our temperament.

THE BALTIC SEA IS DARK as an unlit mirror. Wozniak and I walk along the shore, passing the wooden snack stands that won't be open until spring, and he tells me how he manages his life. He's married, and his wife shares his lifestyle. They swim together in winter, and though Polish is their native language, they communicate in English, which she learned with SuperMemo. Wozniak's days are blocked into distinct periods: a creative period, a reading and studying period, an exercise period, an eating period, a resting period, and then a second creative period. He doesn't get up at a regular hour and is passionate against alarm clocks. If excitement over his research leads him to work into the night, he simply shifts to sleeping in the day. When he sits down for a session of incremental reading, he attends to whatever automatically appears on his computer screen, stopping the instant his mind begins to drift or his comprehension falls too low and then moving on to the next item in the queue. SuperMemo graphs a distribution of priorities that he can adjust as he goes. When he encounters a passage that he thinks he'll need to remember, he marks it; then it goes into a pattern of spaced repetition, and the information it contains will stay in his brain indefinitely.

"Once you get the snippets you need," Wozniak says, "your books disappear. They gradually evaporate. They have been translated into knowledge."

As a science fiction fan, I had always assumed that when computers supplemented our intelligence, it would be because we outsourced some of our memory to them. We would ask questions, and our machines would give oracular—or supremely practical—replies. Wozniak has discovered a different route. When he entrusts his mental life to a machine, it is not to throw off the burden of thought but to make his mind more swift. Extreme knowledge is not something for which he programs a computer but for which his computer is programming him.

I've already told Wozniak that I am not optimistic about my ability to tame old reading habits in the name of optimized knowledge. Books, for me, are not merely sources of information I might want to load into memory but also subjective companions, almost substitute people, and I don't see why I would want to hold on to them in fragments. Still, I tell him I would like to give it a shot.

"So you believe in trying things for yourself?" he asks.

"Yes."

This provides his opening. "In that case, let's go swimming."

At the edge of the sea, I become afraid. I'm a strong swimmer, but there's something about standing on the beach in the type of minuscule bathing suit you get at the gift shop of a discount resort in Eastern Europe, and watching people stride past in their down parkas, that smacks of danger.

"I'm already happy with anticipation," Wozniak says.

"Will I have a heart attack?"

"There is less risk than on your drive here," he answers.

I realize he must be correct. Poland has few freeways, and in the rural north, lines of cars jockey behind communist-era farm machinery until they defy the odds and try to pass. There are spectacular wrecks. Wozniak gives close attention to the qualitative estimate of fatal risks. By graphing the acquisition of knowledge in SuperMemo, he has realized that in a single lifetime one can acquire only a few million new items. This is the absolute limit on intellectual achievement defined by death. So he guards his health. He rarely gets in a car. The Germans on the beach are staring at me. I dive in.

Philosopher William James once wrote that mental life is controlled by noticing. Climbing out of the sea and onto the windy beach, my skin purple and my mind in a reverie provoked by shock, I find myself thinking of a checklist Wozniak wrote a few years ago describing how to become a genius. His advice was straightforward yet strangely terrible: You must clarify your goals, gain knowledge through spaced repetition, preserve health, work steadily, minimize stress, refuse interruption, and never resist sleep when tired. This should lead to radically improved intelligence and creativity. The only cost: turning your back on every convention of social life. It is a severe prescription. And yet now, as I grin broadly and wave to the gawkers, it occurs to me that the cold rationality of his approach may be only a surface feature and that, when linked to genuine rewards, even the chilliest of systems can have a certain visceral appeal. By projecting the achievement of extreme memory back along the forgetting curve, by provably linking the distant future—when we will know so much—to the few minutes we devote to studying today, Wozniak has found a way to condition his temperament along with his memory. He is making the future noticeable. He is trying not just to learn many things but to warm the process of learning itself with a draft of utopian ecstasy.

JOHN SEABROOK

Hello, HAL

FROM *THE NEW YORKER*

Automated communication systems are becoming more common, as anyone calling customer service at a large company can attest. John Seabrook shows that despite the progress that has been made, we are a long way from HAL, *the intelligent computer of 2001.*

NOT LONG AGO, A CALLER DIALLED THE TOLL-FREE number of an energy company to inquire about his bill. He reached an interactive-voice-response system, or I.V.R.—the automated service you get whenever you dial a utility or an airline or any other big American company. I.V.R.s are the speaking tube that connects corporate America to its clients. Companies profess boundless interest in their customers, but they don't want to pay an employee to talk to a caller if they can avoid it; the average human-to-human call costs the company at least five dollars. Once an I.V.R. has been paid for, however, a human-to-I.V.R. call costs virtually nothing.

"If you have an emergency, press one," the utility company's I.V.R. said. "To use our automated services or to pay by phone, press two."

The caller punched two, and was instructed to enter his account number, which he did. An alert had been placed on the account because of a missed payment. "Please hold," the I.V.R. said. "Your call is being transferred to a service representative." This statement was followed by one of the most commonly heard sentences in the English language: "Your call may be monitored."

In fact, the call *was* being monitored, and I listened to it some months later, in the offices of B.B.N. Technologies, a sixty-year-old company, in Cambridge, Massachusetts. Joe Alwan, a vice-president and the general manager of the division that makes B.B.N.'s "caller-experience analytics" software, which is called Avoke, was showing me how the technology can automatically create a log of events in a call, render the speech as text, and make it searchable.

Alwan, a compact man with scrunched-together features who has been at B.B.N. for two years, spoke rapidly but smoothly, with a droll delivery. He projected a graphic of the voice onto a screen at one end of the room. "Anger's the big one," he said. Companies can use Avoke to determine when their callers are getting angry, so that they can improve their I.V.R.s.

The agent came on the line, said his name was Eric, and asked the caller to explain his problem. Eric had a slight Indian accent and spoke in a high, clear voice. He probably worked at a call center in Bangalore for a few dollars an hour, although his pay was likely based on how efficiently he could process the calls. "The company doesn't want to spend more money on the call, because it's a cost," Alwan said. The caller's voice gave the impression that he was white (particularly the way he pronounced the "u" in "*duuude*") and youthful, around thirty:

CALLER: Hey, what's going on is, ah, I got a return-payment notice, right?

AGENT: Mhm.

CALLER: And I checked with my bank, and my bank was saying, well, it didn't even get to you they didn't reject it. So then I was just, like, what's the issue, and then, ah, you guys charge to pay over the phone, so that's why it's not done over the phone, so that's why I do it on the Internet, so—

AGENT: O.K.

CALLER: So I don't . . . know what's going on.

The caller sounded relaxed, but if you listened closely you could hear his voice welling with quiet anger.

The agent quickly looked up the man's record and discovered that he had typed in his account number incorrectly. The caller accepted the agent's explanation but thought he shouldn't be liable for the returned-payment charge. He said, "There's nothing that can be done with that return fee, dude?" The agent explained that another company had levied the charge, but the caller took no notice. "I mean, I would be paying it over the phone, so you guys wanna charge people for paying over the phone, and I'll be—"

People express anger in two different ways. There's "cold" anger, in which words may be overarticulated but spoken softly, and "hot" anger, in which voices are louder and pitched higher. At first, the caller's anger was cold:

AGENT: O.K., sir. I'm gonna go ahead and explain this. . . . O.K., so on the information that you put this last time it was incorrect, so I apologize that you put it incorrectly on the site.

CALLER: O.K., we got past that, bro. So tell me something I don't know. . . .

AGENT: Let's see . . . uh . . . um.

CALLER: Dude, I don't care what company it is. It's your company using that company, so you guys charge it. So you guys should be waiving that shit-over-the-phone shit, pay by phone.

AGENT: But why don't you talk to somebody else, sir. One moment.

By now, the caller's anger was hot. He was put on hold, but B.B.N. was still listening:

CALLER: Motherfucker, I swear. You fucking pussy, you probably don't even have me on hold, you little fucked-up dick. You're gonna wait a long time, bro. You little bitch, I'll fucking find out who you are, you little fucking ho.

After thirty seconds, we could hear bubbling noises—a bong, Alwan thought—and then coughing. Not long afterward, the caller hung up.

THIS SPRING MARKED THE FORTIETH anniversary of HAL, the conversational computer that was brought to life on the screen by Stanley Kubrick and Arthur C. Clarke, in *2001: A Space Odyssey.* HAL has a calm, empathic voice—a voice that is warmer than the voices of the humans in the movie, which are oddly stilted and false. HAL says that he became operational in Urbana, Illinois, in 1992, and offers to sing a song. HAL not only speaks perfectly; he seems to understand perfectly, too. I was a nine-year-old nerd in the making when the film came out, in 1968, and I've been waiting for a computer to talk to ever since—a fantasy shared by many computer geeks. Bill Gates has been touting speech recognition as the next big thing in computing for at least a decade. By giving computers the ability to understand speech, humankind would marry its two greatest technologies: language and toolmaking. To believers, this union can only be a matter of time.

Forty years after *2001,* how close are we to talking to computers?

Today, you can use your voice to buy airplane tickets, transfer money, and get a prescription filled. If you don't want to type, you can use one of the current crop of dictation programs to transcribe your speech; these have been improving steadily and now work reasonably well. If you are driving a car with an onboard navigator, you can get directions in one of dozens of different voices, according to your preference. In a car equipped with Sync—a collaboration of Ford, Microsoft, and Nuance, the largest speech-technology company in the world—you can use your voice to place a phone call or to control your iPod, both of which are useful when you are in what's known in the speech-recognition industry as "hands-busy, eyes-busy" situations. State-of-the-art I.V.R.s, such as Google's voice-based 411 service, offer natural-language understanding—you can speak almost as you would to a human operator, as opposed to having to choose from a set menu of options. I.V.R. designers create vocal personas like Julie, the perky voice that answers Amtrak's 800 number; these voices can be "tuned" according to a company's branding needs. Calling Virgin Mobile gets you a sassy-voiced young woman, who sounds as if she's got her feet up on her desk.

Still, these applications of speech technology, useful though they can be, are a far cry from HAL—a conversational computer. Computers still flunk the famous Turing Test, devised by the British mathematician Alan Turing, in which a computer tries to fool a person into thinking that it's human. And, even within limited applications, speech recognition never seems to work as well as it should. North Americans spent forty-three billion minutes on the line with an I.V.R. in 2007; according to one study, only one caller in ten was satisfied with the experience. Some companies have decided to switch back to touch-tone menus, after finding that customers prefer pushing buttons to using their voices, especially when they are inputting private information, such as account numbers. Leopard, Apple's new operating system for the Mac, responds to voice commands, which is wonderful for people with handicaps and disabilities but extremely annoying if you have to listen to Alex, its computer-generated voice, converse with a co-worker all day.

Roger Schank was a twenty-two-year-old graduate student when *2001* was released. He came toward the end of what today appears to have been a golden era of programmer-philosophers—men like Marvin Minsky and Seymour Papert, who, in establishing the field of artificial intelligence, inspired researchers to create machines with human intelligence. Schank has spent his career trying to make computers simulate human memory and learning. When he was young, he was certain that a conversational computer would eventually be invented. Today, he's less sure. What changed his thinking? Two things, Schank told me: "One was realizing that a lot of human speech is just chatting." Computers proved to be very good at tasks that humans find difficult, like calculating large sums quickly and beating grand masters at chess, but they were wretched at this, one of the simplest of human activities. The other reason, as Schank explained, was that "we just didn't know how complicated speech was until we tried to model it." Just as sending men to the moon yielded many fundamental insights into the nature of space, so the problem of making conversational machines has taught scientists a great deal about how we hear and speak. As the Harvard cognitive scientist Steven Pinker wrote to me, "The consensus as far as I have experienced it among A.I. researchers is that natural-language processing is extraordinarily difficult, as it could involve the entirety of a person's knowledge, which of course is extraordinarily difficult to model on a computer."

After fifty years of research, we aren't even close.

SPEECH BEGINS WITH A PUFF of breath. The diaphragm pushes air up from the lungs, and this passes between two small membranes in the upper windpipe, known as the vocal folds, which vibrate and transform the breath into sound waves. The waves strike hard surfaces inside the head—teeth, bone, the palate. By changing the shape of the mouth and the position of the tongue, the speaker makes vowels and consonants and gives timbre, tone, and color to the sound.

That process, being mechanical, is not difficult to model, and, indeed, humans had been trying to make talking machines long before A.I. existed. In the late eighteenth century, a Hungarian inventor named Wolfgang von Kempelen built a speaking machine by modelling the human vocal tract, using a bellows for lungs, a reed from a bagpipe for the vocal folds, and a keyboard to manipulate the "mouth." By playing the keys, an operator could form complete phrases in several different languages. In the nineteenth century, Kempelen's machine was improved on by Sir Charles Wheatstone, and that contraption, which was exhibited in London, was seen by the young Alexander Graham Bell. It inspired him to try to create his own devices, in the hope of allowing non-hearing people (Bell's mother and his wife were deaf) to speak normally. He didn't succeed, but his early efforts led to the invention of the telephone.

In the twentieth century, researchers created electronic talking machines. The first, called the Voder, was engineered by Bell Labs—the famed research division of AT&T—and exhibited at the 1939 World's Fair, in New York. Instead of a mechanical system made of a reed and bellows, the Voder generated sounds with electricity; as with Kempelen's speaking machine, a human manipulated keys to produce words. The mechanical-sounding voice became a familiar attribute of movie robots in the nineteen-fifties (and, later, similar synthetic-voice effects were a staple of nineteen-seventies progressive rock). In the early sixties, Bell Labs programmed a computer to sing "Daisy, Daisy, give me your answer do." Arthur C. Clarke, who visited the lab, heard the machine sing, and he and Kubrick subsequently used the same song in HAL's death scene.

Hearing is more complicated to model than talking, because it involves signal processing: converting sound from waves of air into electrical impulses. The fleshy part of the ear and the ear canal capture sound waves and direct them to the eardrum, which vibrates as it is struck. These vibrations then push on the ossicles, which form a three-boned lever—that Rube Goldbergian contraption of the middle ear—that helps amplify the sound. The impulses pass into the fluid of the cochlea, which is lined with tiny hairs called cilia. They translate the impulses into electrical signals, which then travel along neural pathways to the brain. Once signals reach the brain, they are "recognized," either by associative memories or by a rules-based system—or, as Pinker has argued, by some combination of the two.

The human ear is exquisitely sensitive; research has shown, for example, that people can distinguish between hot and cold coffee simply by hearing it poured. The ear is particularly attentive to the human voice. We can differentiate among different voices speaking together, and we can isolate voices in the midst of traffic and loud music, and we can tell the direction from which a voice is coming—all of which are difficult for computers to do. We can hear smiles at the other end of a telephone call; the ear recognizes the sound variations caused by the spreading of the lips. That's why call-center workers are told to smile no matter what kind of abuse they're taking.

THE FIRST ATTEMPTS AT SPEECH recognition were made in the nineteen-fifties and sixties, when the A.I. pioneers tried to simulate the way the human mind apprehends language. But where do you start? Even a simple concept like "yes" might be expressed in dozens of different ways—including "yes," "ya," "yup," "yeah," "yeayuh," "yeppers," "yessirree," "aye, aye," "mmmhmm," "uh-huh," "sure," "totally," "certainly," "indeed," "affirmative," "fine," "definitely," "you bet," "you betcha," "no problemo," and "okeydoke"—and what's the rule in that? At Nuance, whose headquarters are outside Boston, speech engineers try to anticipate all the different ways people might say yes, but they still get surprised. For example, designers found that Southerners had more trouble using the system than Northerners did, because when instructed to answer "yes" or "no" Southerners regularly added "ma'am" or "sir," depending on the I.V.R.'s gender, and the computer wasn't programmed to recognize that. Also, language isn't static; the rules change. Researchers taught machines that when the pitch of a voice rises at the end of a sentence it usually means a question, only to have their work spoiled by the emergence of what linguists call "uptalk"—that Valley Girl way of making a declarative sentence sound like a question?—which is now ubiquitous across the United States.

In the seventies and eighties, many speech researchers gradually moved away from efforts to determine the rules of language and took a probabilistic approach to speech recognition. Statistical "learning algorithms"—methods of constructing models from streams of data—were the wheel on which the back of the A.I. culture was broken. As David Nahamoo, the chief technology officer for speech at I.B.M.'s Thomas J. Watson Research Center, told me, "Brute-force computing, based on probability algorithms, won out over the rule-based approach." A speech recognizer, by learning the relative frequency with which particular words occur, both by themselves and within the context of other words, could be "trained" to make educated guesses. Such a system wouldn't be able to understand what words mean, but, given enough data and computing power, it might work in certain, limited vocabulary situations, like medical transcription, and it might be able to perform machine translation with a high degree of accuracy.

In 1969, John Pierce, a prominent member of the staff of Bell Labs, argued in an influential letter to the *Journal of the Acoustical Society of America*, entitled "Whither Speech Recognition," that there was little point in making machines that had speech recognition but no speech understanding. Regardless of the sophistication of the algorithms, the machine would still be a modern version of Kempelen's talking head—a gimmick. But the majority of researchers felt that the narrow promise of speech recognition was better than nothing.

In 1971, the Defense Department's Advanced Research Projects

Agency made a five-year commitment to funding speech recognition. Four institutions—B.B.N., I.B.M., Stanford Research Institute, and Carnegie Mellon University—were selected as contractors, and each was given the same guidelines for developing a speech recognizer with a thousand-word vocabulary. Subsequently, additional projects were funded that might be useful to the military. One was straight out of *Star Trek*: a handheld device that could automatically translate spoken words into other languages. Another was software that could read foreign news media and render them into English.

In addition to DARPA, funding for speech recognition came from telephone companies—principally at Bell Labs—and computer companies, most notably I.B.M. The phone companies wanted voice-based automated calling, and the computer companies wanted a voice-based computer interface and automated dictation, which was a "holy grail project" (a favorite phrase of the industry). But devising a speech recognizer that worked consistently and accurately in real-world situations proved to be much harder than anyone had anticipated. It wasn't until the early nineties that companies finally began to bring products to the consumer marketplace, but these products rarely worked as advertised. The fledgling industry went through a tumultuous period. One industry leader, Lernout & Hauspie, flamed out, in a spectacular accounting scandal.

WHETHER ITS PROVENANCE IS ACADEMIC or corporate, speech-recognition research is heavily dependent on the size of the data sample, or "corpus"—the sheer volume of speech you work with. The larger your corpus, the more data you can feed to the learning algorithms and the better the guesses they can make. I.B.M. collects speech not only in the lab and from broadcasts but also in the field. Andy Aaron, who works at the Watson Research Center, has spent many hours recording people driving or sitting in the front seats of cars in an effort to develop accurate speech models for automotive commands. That's because, he told me, "when people speak in cars they don't speak the same way they do in an office." For example, we talk more loudly in cars, because of a phenomenon known as the Lombard effect—the speaker involuntarily raises his voice to compensate for background noise. Aaron collects speech both for recognizers and for synthesizers—computer-generated voices. "Recording for the recognizer and for the synthesizer couldn't be more different," he said. "In the case of the recognizer, you are teaching the system to correctly identify an unknown speech sound. So you feed it lots and lots of different samples, so that it knows all the different ways Americans might say the phoneme 'oo.' A synthesizer is the opposite. You audition many professional speakers and carefully choose one, because you like the sound of his voice. Then you record that speaker for dozens of hours, saying sentences that contain many diverse combinations of phonemes and common words."

B.B.N. came to speech recognition through its origins as an acoustical engineering firm. It worked on the design of Lincoln Center's Philharmonic Hall in the mid-sixties, and did early research in measuring noise levels at airports, which led to quieter airplane engines. In 1997, B.B.N. was bought by G.T.E., which subsequently merged with Bell Atlantic to form Verizon. In 2004, a group of B.B.N. executives and investors put together a buyout, and the company became independent again. The speech they use to train their recognizers comes from a shared bank, the Linguistic Data Consortium.

During my visit to Cambridge, I watched as a speech engine transcribed a live Al Jazeera broadcast into more or less readable English text, with only a three-minute lag time. In another demo, software captured speech from podcasts and YouTube videos and converted it into text, with impressive accuracy—a technology that promises to make video and audio as easily searchable as text. Both technologies are now available commercially, in B.B.N.'s Broadcast Monitoring System and in EveryZing, its audio-and-video search engine. I also saw B.B.N.'s English-to-Iraqi Arabic translator; I had seen I.B.M.'s, known as the Multilingual Automatic Speech-to-Speech Translator, or MASTOR, the week before. Both worked amazingly well. At I.B.M., an English speaker made a comment ("We are here to provide humanitarian assistance for your town") to an Iraqi. The machine repeated his sentence in English, to make sure it was understood. The MASTOR then translated the sentence into Arabic and said it out loud. The Iraqi answered in Arabic; the machine repeated the sentence in Arabic and then delivered it in English. The entire exchange took about five seconds, and combined state-of-the-art speech recognition, voice synthesis, and machine translation. Granted, the conversation was limited to what you might discuss at a checkpoint in Iraq. Still, for what they are, these translators are triumphs of the statistics-based approach.

What's missing from all these programs, however, is emotional recognition. The current technology can capture neither the play of emphasis, rhythm, and intonation in spoken language (which linguists call prosody) nor the emotional experience of speaking and understanding language. Descartes favored a division between reason and emotion, and considered language to be a vehicle of the former. But speech without emotion, it turns out, isn't really speech. Cognitively, the words should mean the same thing, regardless of their emotional content. But they don't.

SPEECH RECOGNITION IS A MULTIDISCIPLINARY field, involving linguists, psychologists, phoneticians, acousticians, computer scientists, and engineers. At speech conferences these days, emotional recognition is a hot topic. Julia Hirschberg, a professor of computer science at Columbia University, told me that at the last prosody conference she attended "it seemed like three-quarters of the presentations were on emotional recognition." Research is focussed both on how to recognize a speaker's emotional state and on how to make synthetic voices more emotionally expressive.

Elizabeth Shriberg, a senior researcher in the speech group at S.R.I. International (formerly Stanford Research Institute), said,

"Especially when you talk about emotional speech, there is a big difference between acted speech and real speech." Real anger, she went on, often builds over a number of utterances, and is much more variable than acted anger. For more accurate emotional recognition, Shriberg said, "we need the kind of data that you get from 911 and directory-assistance calls. But you can't use those, for privacy reasons, and because they're proprietary."

At SAIL—the Speech Analysis and Interpretation Laboratory, on the campus of the University of Southern California, in Los Angeles—researchers work mostly with scripted speech, which students collect from actors in the U.S.C. film and drama programs. Shrikanth Narayanan, who runs the lab, is an electrical engineer, and the students in his emotion-research group are mainly engineers and computer scientists. One student was studying what happens when a speaker's face and voice convey conflicting emotions. Another was researching how emotional states affect the way people move their heads when they talk. The research itself can be a grind. Students painstakingly listen to voices expressing many different kinds of emotion and tag each sample with information, such as how energetic the voice is and its "valence" (whether it is a negative or a positive emotion). Anger and elation are examples of emotions that have different valences but similar energy; humans use context, as well as facial and vocal cues, to distinguish them. Since the researchers have only the voice to work with, at least three of them are required to listen and decide what the emotion is. Students note voice quality, pacing, language, "disfluencies" (false starts, "um"s), and pitch. They make at least two different data sets, so that they can use separate ones for training the computer and for testing it.

Facial expressions are generally thought to be universal, but so far Narayanan's lab hasn't found that similarly universal vocal cues for emotions are as clearly established. "Emotions aren't discrete," Narayanan said. "They are a continuum, and it isn't clear to any one perceiver where one emotion ends and another begins, so you end up studying not just the speaker but the perceiver." The idea is that if you could train the computer to sense a speaker's emotional state by the sound of his voice, you could also train it to respond in kind—the computer might slow down if it sensed that the speaker was confused, or assume a more soothing tone of voice if it sensed anger. One possible application of such technology would be video games, which could automatically adapt to a player's level based on the stress in his voice. Narayanan also mentioned simulations—such as the computer-game-like training exercises that many companies now use to prepare workers for a job. "The program would sense from your voice if you are overconfident, or when you are feeling frustrated, and adjust accordingly," he said. That reminded me of the moment in the novel *2001* when HAL, after discovering that the astronauts have doubts about him, decides to kill them. While struggling with one of the astronauts, Dave, for control of the ship, HAL says, "I can tell from your voice harmonics, Dave, that you're badly upset. Why don't you take a stress pill and get some rest?"

But, apart from call-center voice analytics, it's hard to find many credible applications of emotional recognition, and it is possible that true emotional recognition is beyond the limits of the probabilistic approach. There are futuristic projects aimed at making emotionally responsive robots, and there are plans to use such robots in the care of children and the elderly. "But this is very long-range, obviously," Narayanan said. In the meantime, we are going to be dealing with emotionless machines.

There is a small market for voice-based lie detectors, which are becoming a popular tool in police stations around the country. Many are made by Nemesysco, an Israeli company, using a technique called "layered voice analysis" to analyze some hundred and thirty parameters in the voice to establish the speaker's psychological state. The academic world is skeptical of voice-based lie detection, because Nemesysco will not release the algorithms on which its program is based; after all, they are proprietary. Layered voice analysis has failed in two independent tests. Nemesysco's American distributor says that's because the tests were poorly designed. (The company played

Roger Clemens's recent congressional testimony for me through its software, so that I could see for myself the Rocket's stress levels leaping.) Nevertheless, according to the distributor more than a thousand copies of the software have been sold—at fourteen thousand five hundred dollars each—to law-enforcement agencies and, more recently, to insurance companies, which are using them in fraud detection.

One of the most fully realized applications of emotional recognition that I am aware of is the aggression-detection system developed by Sound Intelligence, which has been deployed in Rotterdam and Amsterdam, and other cities in the Netherlands. It has also been installed in the English city of Coventry, and is being tested in London and Manchester. One of the designers, Peter van Hengel, explained to me that the idea grew out of a project at the University of Groningen, which simulated the workings of the inner ear with computer models. "A colleague of mine applied the same inner-ear model to trying to recognize speech amid noise," he said, "and found that it could be used to select the parts belonging to the speech and leave out the noise." They founded Sound Intelligence in 2000, initially focussing on speech-noise separation for automatic speech recognition, with a sideline in the analysis of non-speech sounds. In 2003, the company was approached by the Dutch national railroad, which wanted to be able to detect several kinds of sound that might indicate trouble in stations and on trains (glass-breaking, graffiti-spraying, and aggressive voices). This project developed into an aggression-detection system based on the sound of people shouting: the machine detects the overstressing of the vocal cords, which occurs only in real aggression. (That's one reason actors only approximate anger; the real thing can damage the voice.)

The city of Groningen has installed an aggression-detector at a busy intersection in an area full of pubs. Elevated microphones spaced thirty metres apart run along both sides of the street, joining an existing network of cameras. These connect to a computer at the police station in Groningen. If the system hears certain sound patterns that correspond with aggression, it sends an alert to the police station, where the police can assess the situation by examining closed-circuit monitors: if necessary, officers are dispatched to the scene. This is no HAL, either, but the system is promising, because it does not pretend to be more intelligent than it is.

I thought the problem with the technology would be false positives—too many loud noises that the machine mistook for aggression. But in Groningen, at least, the problem has been just the opposite. "Groningen is the safest city in Holland," van Hengel said, ruefully. "There is virtually no crime. We don't have enough aggression to train the system properly."

CATHERINE PRICE

The Anonymity Experiment

FROM *POPULAR SCIENCE*

In this interconnected, digital world of ours, going off the grid is not easy, as Catherine Price discovers when she tries it for herself.

IN 2006, DAVID HOLTZMAN DECIDED TO DO AN EXPERIment. Holtzman, a security consultant and former intelligence analyst, was working on a book about privacy, and he wanted to see how much he could find out about himself from sources available to any tenacious stalker. So he did background checks. He pulled his credit file. He looked at Amazon.com transactions and his credit-card and telephone bills. He got his DNA analyzed and kept a log of all the people he called and e-mailed, along with the Web sites he visited. When he put the information together, he was able to discover so much about himself—from detailed financial information to the fact that he was circumcised—that his publisher, concerned about his privacy, didn't let him include it all in the book.

I'm no intelligence analyst, but stories like Holtzman's freak me out. So do statistics like this one: In 2007, 127 million sensitive electronic and paper records (those containing Social Security numbers and the like) were hacked or lost—a nearly 650 percent increase in data breaches from the previous year. In January of the same year, news broke that hackers had stolen somewhere between 45 million and 94 million credit- and debit-card numbers from the databases of the retail company TJX, in one of the biggest data breaches in history. And that November, the British government admitted losing computer discs containing personal data for 25 million people, which is almost half the country's population.

The situation is bad enough that Donald Kerr, the principal deputy director of National Intelligence, proclaimed in an October 2007 speech that "protecting anonymity isn't a fight that can be won." Privacy-minded people have long warned of a world in which an individual's every action leaves a trace, in which corporations and governments can peer at will into your life with a few keystrokes on a computer. Now one of the people in charge of information-gathering for the U.S. government was saying, essentially, that such a world has arrived.

So when [*Popular Science*] suggested I try my own privacy experiment, I eagerly agreed. We decided that I would spend a week trying to be as anonymous as possible while still living a normal life. I would attempt what many believe is now impossible: to hide in plain sight.

A GALLUP POLL OF APPROXIMATELY 1,000 Americans taken in February 1999 found that 70 percent of them believed that the Constitution "guarantees citizens the right to privacy." Wrong. The Constitution doesn't even contain the word.

A number of amendments protect privacy implicitly, as do certain state and federal laws, the most significant of which is the Privacy Act of 1974, which prohibits disclosure of some federal records that contain information about individuals.* Unfortunately, the law is full of exceptions. As Beth Givens, founder and director of the nonprofit Privacy Rights Clearinghouse, put it, the Privacy Act has "so many limitations that it can barely be called a privacy act with a straight face."

In the U.S., unlike other countries, we don't have broad, generally applicable laws to protect our personal information. We've got federal laws that safeguard very specific types of data, like student records, credit reports and DVD rentals. But those have loopholes too.†

In addition, technological advances are quickly rendering many of these laws useless. What good is strong protection for cable records when a technology like TiVo comes along that is not, technically, a "cable service provider"?‡ Or a statute about postal mail in a world where most communication now takes place online? "We're way behind the curve," says Richard Purcell, CEO of the Corporate Privacy Group and former chief privacy officer for Microsoft. "Technology is way ahead of our ability as a society to think about the consequences."

Navigating this technological and legal maze wouldn't be easy; I needed professional help, a privacy guru who could guide me through

*California, where I live, leads the nation in privacy protection. If I'd conducted my experiment elsewhere in the U.S., it would have been even more difficult.

†One oft-cited loophole is in the Driver's Privacy Protection Act of 1994. It was created after a series of crimes linked to Department of Motor Vehicles records, the most notorious of which occurred in 1989: An obsessed fan hired a private investigator to get actress Rebecca Schaeffer's home address from her DMV record and then tracked her down and killed her. Now DMV employees aren't allowed to release personal information. The only problem is that the law has 14 exemptions, including one that allows the release of information to licensed private investigators if they say they're using it for purposes listed in the other 13 exemptions.

‡TiVo actually has a strict privacy policy—but it's the company's own doing. Legally, it's allowed to sell a minute-by-minute record of users' viewing habits.

my week. That man was Chris Jay Hoofnagle, a privacy expert and lawyer who used to run the West Coast office of the Electronic Privacy Information Center (EPIC), a public-interest research center in Washington, D.C., that focuses on privacy and civil-liberties issues.

Hoofnagle had tried his own version of the same thing, partly for fun and partly because of fears of retribution from private investigators he had irritated in his previous job at EPIC. "When moving to San Francisco two years ago, I deliberately gave my new address to no business or government entity," he told me. "As a result, no one really knows where I live." His bills are in aliases, and despite setbacks—like having his power turned off because the company didn't know where to send the statement—he's been successful at concealing his home address.

Now that he's a senior fellow at the University of California at Berkeley's Boalt Hall School of Law, Hoofnagle doesn't keep his office location a secret. So on a sunny afternoon, I set off to meet him there.

Tall and friendly, Hoofnagle has an enthusiastic way of talking about privacy violations that could best be described as "cheerful outrage." He laid out my basic tasks: Pay for everything in cash. Don't use my regular cellphone, landline or e-mail account. Use an anonymizing service to mask my Web surfing. Stay away from government buildings and airports (too many surveillance cameras), and wear a hat and sunglasses to foil cameras I can't avoid. Don't use automatic toll lanes. Get a confetti-cut paper shredder for sensitive documents and junk mail. Sign up for the national do-not-call registry (ignoring, if you can, the irony of revealing your phone number and e-mail address to prevent people from contacting you), and opt out of prescreened credit offers. Don't buy a plane ticket, rent a car, get married, have a baby, purchase land, start a business, go to a casino, use a supermarket loyalty card, or buy nasal decongestant.*

*Pseudoephedrine can be used to make methamphetamine, and thanks to a federal law passed in 2006, your name goes into a log when you buy products that contain it.

By the time I left Hoofnagle's office, a week was beginning to sound like a very long time.

AFTER WITHDRAWING SEVEN DAYS' WORTH of cash, I officially began my experiment by attempting to buy an anonymous cellphone. This was a crucial step. My cellphone company keeps records of every call I make and receive. What's more, my phone itself can give away my location. Since 1999, the Federal Communications Commission has required that all new cellphones in the U.S. use some form of locating technology that makes it possible to find them within 1,000 feet 95 percent of the time; the technology works as long as your phone is on, even if it isn't being used. This is to help 911 responders locate you, but the applications are expanding. Advertisers hope to use cellphone GPS to send text-message coupons to lure you into stores as you pass by, and services such as Loopt and Buddy Beacon allow you to see a map of where your friends are, in real time, using either their cellphone signals or GPS.* I left an outgoing message explaining that I wasn't going to be using my normal cellphone for a week and then turned it off.

Wearing a baseball cap and sunglasses, I walked into an AT&T store and immediately noticed several black half-globes suspended from the ceiling: surveillance cameras. I needed to keep my head down. When I tried to pay for my new phone, the cashier swiped its bar code, looked up at me with her fingers poised above her keyboard, and asked me for identification. "I don't have any on me," I lied.

She seemed mildly annoyed and asked for my name and address.

"I'm sorry," I said, "but I don't really want my information in the system."

"We need your information."

"Why?"

"For billing purposes."

*These services are voluntary, but they illustrate the privacy-killing potential of cellphone GPS.

"But it's a prepaid card. You don't need to bill me."

This, apparently, was irrelevant. "We need to put your information into the system," she said again. "Otherwise you can't buy the phone."

I didn't buy the phone. Instead I walked across the street to a generic cellphone store where a young clerk with pink hair and black-framed glasses was sitting behind the cash register, text messaging. "So do you want me to, like, just put in some random name?" she asked. Before I knew it, she'd christened me Mike Smith, born October 18, 2007.* As she charged minutes to my phone, I overheard a young man next to me tell a different clerk that he wanted to activate a cellphone that was registered under his mother's name. "That's no problem at all," said the clerk. "We just need her Social Security number." Unfazed, the man called his mom. He was dictating the number to the clerk as Mike Smith walked out the door.

MY NEW PHONE WAS ANONYMOUS, but I still needed to be careful. If I didn't want it to be traceable back to me, I had to disguise my outgoing calls and minimize the number of calls that I received; records of both could be used to identify me. I changed the phone's settings so that its number wouldn't show up when I placed calls† and bought a prepaid calling card to use on top of my cellphone. That way, if anyone were to pull a record of my outgoing calls, they would just see the calling-card number.

If masking your cellphone number is difficult, hiding your online activity is nearly impossible. Anytime you access the Internet, your Internet service provider (ISP) knows you're online, and it might soon keep track of more. In 2005, the European Parliament passed

*She'd asked for my birth date to use as an activation code, but it turned out she really just needed any eight-digit series of numbers.

†Because of something called "automatic number identification," there's no way to stop your information from showing up when you call toll-free or 900 numbers.

legislation requiring phone and Internet providers to retain records of calls and online activity for between six months and two years. In 2006, then–U.S. attorney general Alberto Gonzales and FBI director Robert Mueller met privately with America's major ISPs to request that they, too, hold on to these records for two years. Search engines already keep records of queries, a practice that's become enough of a concern among users that in December 2007, Ask.com launched AskEraser, a service that deletes your searches within hours. When you send an unencrypted e-mail, it can be intercepted and read and may be stored indefinitely on a server, even if you've deleted it. And Web sites routinely retain such information as how you got there, how long you lingered on each page, and your scrolling, clicks and mouse-overs.

But I'm getting ahead of myself. My first challenge was to figure out how to connect to the Internet. I subscribed to Anonymizer, a service that uses a technique called secure-shell tunneling to create a virtual link between your computer and Anonymizer's proxy servers (that is, the servers that act as middlemen between your computer and the sites you're visiting). That meant my ISP wouldn't know what Web sites I was visiting, and the sites I visited wouldn't know it was me.

Anonymizer has two potential weaknesses, though. First, Anonymizer itself knows what sites you're visiting, although the company claims not to retain this information. And then there's the conspiracy theory. "There are reports that the government has sneakily had people volunteer to run Anonymizer server nodes who are actually 'quislings'"—traitors—Holtzman told me. "I don't know if it's true," he said. "But if it were my job to spy on people, I'd be doing it."

There are Anonymizer alternatives (the freeware Tor is probably the best-known), but according to Holtzman, if you want to be sure of anonymity, "you just cannot use your own computer. The only way to do so is if it's brand-new and you never put it online." Unfortunately, I didn't have a brand-new computer, and I needed to use the Internet. I decided to avoid using my own ISP whenever possible.

Instead I needed to either piggyback on neighbors' open connections or use public Wi-Fi hotspots.*

Once I got online, I had other challenges, such as the notorious "cookie," a small text file that Web sites leave on your computer so that you can be identified when you return. Cookies often make the Internet easier to use; they remember your login name, for example, and some sites now deny access to visitors who refuse cookies. But if you want to remain anonymous online, you've got to toss your cookies. When I looked through my cookie cache, I found them from all sorts of places, including one from a site called Vegan Porn (it's not as naughty as it sounds) and another, from Budget Rent A Car, that wasn't set to expire until 2075.

Lastly, there was the question of e-mail. I set my usual address to forward to a Hotmail account I'd created with fake user information and signed up for a free account through Hushmail, a service that allows you to send encrypted, anonymous e-mail. I figured that if I monitored my messages through Hotmail but responded using only Hushmail, no one would be able to connect the two accounts—or know definitively that the person checking the Hotmail was me. Only later did I discover that even Hushmail has occasionally spilled information to the feds.

I STARTED MARKING ITEMS OFF Hoofnagle's to-do list. I signed up for the do-not-call registry to avoid telemarketers and sent a letter opting out of all prescreened offers of credit. I called my bank and opted out of its information sharing.† Then I called my phone com-

*Even then, I still wouldn't be entirely anonymous. Every networking device in every computer is assigned a media access control (MAC) address, a unique identifying number picked up by your router when you go online. To learn how to find your MAC address, go to www.dcn.fnal.gov/DCG-Docs/mac/index.html.

†Banks sell lists of information that you'd think would be kept private — transaction histories, bank balances, where you've sent payments—and can continue to do so even if your account is closed. But banks are better than they used to be:

pany and told them I didn't want them to share my CPNI—customer proprietary network information.

Your CPNI includes records of what services you use, what types of calls you make, when you place them, and a log of the numbers you've called. Before 1996, phone companies were allowed to freely sell this information to third parties for marketing purposes. Today, thanks to legislation limiting what they can do without your permission, CPNI is mainly used to sell you other services offered by your phone company, such as a new long-distance plan.

When my phone company's automated system picked up, a voice announced that my call might be monitored or recorded but that I could ask to be on an unrecorded line. So I did. "Uh, OK," the representative said. "But all the lines are recorded automatically. If you don't want to be recorded, I'm going to have to call you back."

"How long will that take?" I asked, having already spent 10 minutes on hold.

"We'll call you as soon as we have a chance," he said. "Probably within an hour." In other words, the cost of privacy would be an hour of my time.

I told him that a recorded line was fine and then asked him to stop using my CPNI to market things to me. He agreed. Then he asked if I had a few minutes to talk about my phone service and proceeded to use my CPNI to try to sell me a unified messaging system.

THIS WAS GETTING EXHAUSTING. I'D thought a yoga class would be a nice break, but I'd forgotten one thing: The yoga studio I go to has a computer system that keeps track of all its students' names. I scrawled "CPrice" illegibly on the sign-in sheet and paid in cash. I thought I'd gotten away with my ploy until the end of class when, just after our final "om," the teacher picked up a piece of paper

---

Until the Gramm-Leach-Bliley Act in 1999, banks could even sell account and credit-card numbers to unaffiliated third parties.

that the front desk had slipped under the door. "Would whoever signed in as number 19 please stop by the front on the way out?" he asked. "They couldn't read your signature."

I doubt the young Buddhists behind the yoga-studio desk are profit-minded enough to sell my personal information, but many other businesses are. Data broker Web sites sell lists of information you never thought would be for sale—records of 750,000 people who signed up for medical alert services, for example, or a list of 11,418 people, mostly men over the age of 55, who bought a particular herbal sexual-potency product in September or October. Private investigators buy phone records from pizza-delivery places, and a few years ago, data aggregator LexisNexis advertised that it, too, used pizza-delivery records to get hard-to-find phone numbers. If you want to invalidate some of the information on the lists, you could move, but you'd have to carry your own boxes—moving companies sell lists of new addresses to marketers.

More disturbing is the fact that this relatively disparate information is frequently rounded up by other data-aggregator companies such as ChoicePoint and Acxiom. Acxiom's databases contain records on 96 percent of American households. Its newest customer intelligence database, InfoBase-X, includes 199 million names and can draw on 1,500 "data elements" to help companies market to potential customers, including "Life Event, Buying Activity, Travel, Behavior, Ethnicity, Lifestyle/Interests, Real Property, Automotive and more."

These companies are minimally regulated, in part because the government itself is one of their largest clients. Contracting data-collection projects to outside companies allows the government to purchase data that would be illegal for it to collect itself. Take, for example, what happened in 2002 when a now-defunct information-mining company and Department of Defense contractor called Torch Concepts got five million itinerary records for JetBlue passengers—records that included names, addresses and phone numbers—for a project whose goal was ostensibly to identify high-risk airline passengers. Torch Concepts then bought demographic data from Acxiom on about 40 percent of the passengers whose records JetBlue had released.

This demographic data included passengers' genders, homeownership status, occupations, length of time spent at their residences, income level, vehicle information, Social Security numbers and how many kids they had. The company used the information to create detailed profiles of the passengers, including one (with the name stripped off but all other information still intact) that it used as part of a presentation to pitch potential clients.

TRANSPORTATION WAS TRICKY. I'D BEEN wearing my hat and sunglasses so I couldn't be recognized on cameras, but to take buses or the train would be to willingly subject myself to heavy surveillance, and that was against my rules. I couldn't drive my car through toll plazas—they're covered in cameras, and if you have an automatic toll-payment system that uses a pre-paid account, like E-ZPass or, in the Bay Area, FasTrak, you leave behind a record.*

I'd also learned about EDRs, or event data recorders, small devices installed in most new passenger vehicles that monitor things like speed, steering-wheel angle, acceleration, braking and seatbelt use. EDRs were first developed in the 1970s and began to be installed as part of airbag systems in the 1990s.† If safety sensors in your car detect a sudden deceleration, they trigger the airbag, and the EDR

*Some states have sensors along the road that use toll passes to identify cars as they pass through two points. This information is used to make calculations about traffic speed and feed electronic billboards that provide up-to-the-minute estimated driving times to various locations. This information could also be used, hypothetically, to automatically issue speeding tickets.

†If you want to see whether your car has an EDR, check your owner's manual—it's usually disclosed in the section about airbags. But EDRs aren't the only thing to be aware of. Car-rental companies have used GPS to tell when customers violated the terms of their contracts by speeding or crossing state lines.

retains a record of what happened in the seconds preceding and following the collision.

But today, EDRs are part of sophisticated systems that do much more. If you subscribe to GM's OnStar service, for example, and get in a wreck, your car will notify OnStar so a representative can contact you through the speaker system in your car and medics can respond to the scene more quickly.

It's hard to complain about a voluntary service that could save your life, but other features are more intrusive. Starting in 2009, OnStar will be able to remotely deactivate a car's accelerator, forcing it to drive at a top speed of five miles an hour—which is great if your car is stolen but not so good if someone were to hack into OnStar's computers. Plus, systems like these include a two-way microphone and speakers that the company can activate remotely, which means they can be used for eavesdropping.

The FBI took advantage of this capability a few years ago, when it got court authority to compel a company (which was unnamed in court documents) to turn on the microphone in a suspect's car to monitor conversations. The FBI eventually lost the case on appeal, but for reasons relating to safety, not privacy: a court decided that the agency had forced the company to breach its contract with the suspect because using the car's microphone for surveillance rendered it useless in case of emergency.

Fortunately, my car is old enough that it doesn't have an EDR. If I were to just drive around my neighborhood, I'd only have to worry about traffic and red-light cameras, whose images generally aren't archived unless something noteworthy happens. But I needed to go to San Francisco—the International Association of Privacy Professionals was having a conference. The problem was that attending it would require getting across the Bay Bridge.

At first I thought this might be impossible. Then I remembered Casual Carpool, an informal system in which drivers can use toll-free lanes by picking up passengers throughout the East Bay and dropping them off in San Francisco.

Up to that point, I'd been wearing a cap and sunglasses every time

I went outside.* I liked my camouflage. It made me feel like I could be mistaken for J. Lo. But I thought that for my grand trip into surveillance-camera-dense San Francisco, I should try something different. I decided to wear my visor.

Let me be clear: This was no ordinary face visor. Designed to provide complete sun protection, it was more of a mask, with a wraparound piece of dark plastic that extended from my forehead all the way down to my chin. It made me look like a welder. It also made it difficult to see. But I still managed to find a car, and surprisingly, no one commented on the visor. In fact, they didn't talk to me at all.

At the conference, I switched from visor to hat, which made it easier to blend into the crowd of more than 900 "privacy professionals" (whose existence is a sign that more companies are taking privacy seriously). Here I listened to lectures on two technologies—IPv6 and RFID—that have significant privacy implications. IPv6, which stands for Internet Protocol Version 6 and is an eventual replacement for our current Internet protocol, IPv4, would allow mobile devices to connect to one another directly without any need for a server; it also means that your camera, PDA or just about any other gadget could have a unique identifier that would make it possible to track you in real time.

And then there are RFID (radio-frequency identification) chips, small devices that consist of a microchip and an antenna that use radio waves to identify objects and people.† About five years ago,

*The quality of the images taken by most surveillance cameras—at least the surveillance cameras of today—is unrefined enough that you don't need too much of a disguise.

†Starting in 2007, all new U.S. passports are embedded with RFID chips that contain the person's identifying information and a photo, and research is under way on how to embed the chips in paper currency. RFID tags are already used to "microchip" pets. One company, VeriChip, has implanted 500 people in the U.S. with RFID chips and it has proposed replacing military dog tags by implanting the chips into American soldiers. It sounds far-fetched, but this is a real enough possibility that last October, California governor Arnold Schwarzenegger signed a bill forbidding employers to force employees to have RFID chips implanted under their skin.

these chips (often called tags) were the obsession of conspiracy theorists everywhere. But the time to really worry about RFID may be near. Experts like Holtzman predict that soon the price of the tags will drop enough that they will be attached to almost everything we buy and will become so small as to basically be invisible. "You couldn't get away with this experiment in a couple years because of the RFID chips," Holtzman told me later. "You'd literally have to get rid of everything you own and start over, since every artifact you'd bought from a major manufacturer would probably have a chip embedded in it that could identify you as the buyer."

Just before the conference ended, I tracked down Richard Purcell, the former CPO of Microsoft. After dodging security cameras in the hallway, we ducked into an empty ballroom to talk. He was not encouraging. "The thing is, surveillance is a fact of our electronic society," he said. "You are going to be tracked. One has to be thoughtful about that." He's right. No one knows exactly how many surveillance cameras are being used in the U.S. right now, but consider that the much-smaller U.K. has three to four million.

And more cameras arrive all the time. The New York City Police Department, for instance, aims to install an additional 3,000 public and private security cameras below Canal Street, with video feeds that could broadcast directly to the Department of Homeland Security and the FBI. That's understandable—once the Freedom Tower goes up at the World Trade Center site, lower Manhattan will once again be home to one of the most conspicuous terrorist targets in the world. But the surveillance-camera craze has begun to veer into absurdity: The British government recently approved funding to pay for cameras in the hats of more than 2,000 police officers.

THE PROBLEM WITH CASUAL CARPOOL is that it primarily runs into the city, which left me without a way to get home. I decided to take a cab but then noticed a plastic decal that read "Smile, you're on camera!" Whatever. By that point, one more camera was the least of my worries. Instead, I spent the cab ride mulling the most common counterargument to concerns over lost privacy: So what? If giving up personal information makes it easier for me to shop online, so be it. If total surveillance can prevent terrorist attacks, bring on Big Brother.

Here's the thing, though—we don't know what information is being collected about us, with whom it's being shared, what it's being used for, or where it's being held. As companies and the government collect more and more data on us, some of it will inevitably be incorrect, and the effect of those errors could range from trivial to severe. It's not a big deal to get coupons for products you don't want, but if a mistake in your file or an identity theft caused by a data breach drives down your credit score, you could find yourself knocked into the subprime-mortgage market. And privacy-invading safeguards don't just catch bad guys. Anyone could end up like Senator Ted Kennedy, who was erroneously placed on a do-not-fly list because a terrorist had once used the alias "T. Kennedy."

For now, few systems are in place to help us understand what data is being gathered or correct the inevitable mistakes, and in the absence of laws that define punishments for data breaches—and judges who enforce them—companies can walk away from serious privacy violations with nothing more than a slap on the wrist.

Case in point: When EPIC filed a complaint with the FTC against JetBlue for disclosing passenger information to Torch Concepts, the agency never publicly opened an investigation; in response to a separate suit filed by JetBlue passengers, a federal judge agreed that the company had violated its privacy policies but dismissed the lawsuit because passengers weren't able to prove that anything had happened to them as a result of the profiling, and that JetBlue hadn't "unjustly enriched" itself by sharing the information. And because this kind of news is so often met with no more than a collective shrug, such privacy violations are likely to keep happening.

At the end of my week of paranoia, I met Hoofnagle at the Yerba Buena Center of the Arts in San Francisco so he could grade me on my performance. His verdict: I did a pretty good job. But his approval seemed less satisfying when I considered all the aspects of my life that made it easier to minimize my digital trail. I don't use pay-per-view or FasTrak. I don't work in an office, which would require an ID card and logging on to and e-mailing from company computers. I don't use Instant Messenger, play online games, visit chatrooms or sell things on eBay. I've never been married or arrested, so few public records are associated with my name.

Also, spending one week undercover doesn't do anything about the information that's already out there—information that, for the most part, I volunteered. Countless Web sites have records on me. UPS, FedEx, and the Department of Motor Vehicles—not to mention every store from which I've shopped online—know where I live. My bank, credit-card company, gym, health insurance company and phone company all have me in their records, and my information is in alumni databases. Both my college and graduate school have lost laptops containing my Social Security number.

I was reminded of something Holtzman had told me earlier that week. "No matter what you do, you'll never really know if you're successful at keeping private," he said. "There are all sorts of trails you leave that you'll never even know about."

Once Hoofnagle had left, I walked through an exhibit at Yerba Buena, "Dark Matters," that happened to feature—no kidding—pieces about surveillance. One installation in particular captivated me. Called *Listening Post*, it was a darkened room with gray walls, empty except for a large lattice hanging from the ceiling made from 231 small screens, each about the size and shape of a dollar bill. The screens displayed scrolling blue-green sentence fragments that were being culled, real-time, from Internet chatrooms. Occasionally the program would search for sentences that began with key words—"I am," "I like," "I love"—and the results would scroll across the screens. "I love my new cellphone." "I love you and your sexy hair." "I love Quark."

It was strangely calming, standing in this dim room, watching the words and thoughts of strangers reveal themselves to me. I still had my hat on, but for once there were no surveillance cameras, so I sat down on a bench in the room and pulled out my notebook, grateful to finally be the observer rather than the observed. And then, out of the corner of my eye, I saw her: a security guard standing in the room's darkened corner—silent, motionless, watching.

GREGG EASTERBROOK

The Sky Is Falling

FROM *THE ATLANTIC MONTHLY*

Global economic crisis, terrorism, climate change—there's enough to be worried about, right? Gregg Easterbrook reminds us that the threat of a catastrophic asteroid impact on the Earth is still real, and we are not doing enough about it.

BREAKTHROUGH IDEAS HAVE A WAY OF SEEMING OBvious in retrospect, and about a decade ago, a Columbia University geophysicist named Dallas Abbott had a breakthrough idea. She had been pondering the craters left by comets and asteroids that smashed into Earth. Geologists had counted them and concluded that space strikes are rare events and had occurred mainly during the era of primordial mists. But, Abbott realized, this deduction was based on the number of craters found on land—and because 70 percent of Earth's surface is water, wouldn't most space objects hit the sea? So she began searching for underwater craters caused by impacts rather than by other forces, such as volcanoes. What she has found is spine-chilling: evidence that several enormous asteroids or comets have slammed into our planet quite recently, in geologic terms. If Abbott is right, then you may be here today, reading this article, only because by sheer chance those objects struck the ocean rather than land.

Abbott believes that a space object about 300 meters in diameter hit the Gulf of Carpentaria, north of Australia, in 536 A.D. An object that size, striking at up to 50,000 miles per hour, could release as much energy as 1,000 nuclear bombs. Debris, dust, and gases thrown into the atmosphere by the impact would have blocked sunlight, temporarily cooling the planet—and indeed, contemporaneous accounts describe dim skies, cold summers, and poor harvests in 536 and 537. "A most dread portent took place," the Byzantine historian Procopius wrote of 536; the sun "gave forth its light without brightness." Frost reportedly covered China in the summertime. Still, the harm was mitigated by the ocean impact. When a space object strikes land, it kicks up more dust and debris, increasing the global-cooling effect; at the same time, the combination of shock waves and extreme heating at the point of impact generates nitric and nitrous acids, producing rain as corrosive as battery acid. If the Gulf of Carpentaria object were to strike Miami today, most of the city would be leveled, and the atmospheric effects could trigger crop failures around the world.

What's more, the Gulf of Carpentaria object was a skipping stone compared with an object that Abbott thinks whammed into the Indian Ocean near Madagascar some 4,800 years ago, or about 2,800 B.C. Researchers generally assume that a space object a kilometer or more across would cause significant global harm: widespread destruction, severe acid rain, and dust storms that would darken the world's skies for decades. The object that hit the Indian Ocean was three to five kilometers across, Abbott believes, and caused a tsunami in the Pacific 600 feet high—many times higher than the 2004 tsunami that struck Southeast Asia. Ancient texts such as Genesis and the Epic of Gilgamesh support her conjecture, describing an unspeakable planetary flood in roughly the same time period. If the Indian Ocean object were to hit the sea now, many of the world's coastal cities could be flattened. If it were to hit land, much of a continent would be leveled; years of winter and mass starvation would ensue.

At the start of her research, which has sparked much debate among specialists, Abbott reasoned that if colossal asteroids or comets strike the sea with about the same frequency as they strike land, then given the number of known land craters, perhaps 100 large impact craters might lie beneath the oceans. In less than a decade of searching, she and a few colleagues have already found what appear to be 14 large underwater impact sites. That they've found so many so rapidly is hardly reassuring.

Other scientists are making equally unsettling discoveries. Only in the past few decades have astronomers begun to search the nearby skies for objects such as asteroids and comets (for convenience, let's call them "space rocks"). What they are finding suggests that near-Earth space rocks are more numerous than was once thought, and that their orbits may not be as stable as has been assumed. There is also reason to think that space rocks may not even need to reach Earth's surface to cause cataclysmic damage. Our solar system appears to be a far more dangerous place than was previously believed.

The received wisdom about the origins of the solar system goes something like this: the sun and planets formed about 4.5 billion years ago from a swirling nebula containing huge amounts of gas and dust, as well as relatively small amounts of metals and other dense substances released by ancient supernova explosions. The sun is at the center; the denser planets, including Earth, formed in the middle region, along with many asteroids—the small rocky bodies made of material that failed to incorporate into a planet. Farther out are the gas-giant planets, such as Jupiter, plus vast amounts of light elements, which formed comets on the boundary of the solar system. Early on, asteroids existed by the millions; the planets and their satellites were bombarded by constant, furious strikes. The heat and shock waves generated by these impacts regularly sterilized the young Earth. Only after the rain of space objects ceased could life begin; by then, most asteroids had already either hit something or found stable orbits that do not lead toward planets or moons. Asteroids still exist, but most were assumed to be in the asteroid belt, which lies between Mars and Jupiter, far from our blue world.

As for comets, conventional wisdom held that they also bombarded the planets during the early eons. Comets are mostly frozen water mixed with dirt. An ancient deluge of comets may have helped create our oceans; lots of comets hit the moon, too, but there the light elements they were composed of evaporated. As with asteroids, most comets were thought to have smashed into something long ago; and, because the solar system is largely void, researchers deemed it statistically improbable that those remaining would cross the paths of planets.

These standard assumptions—that remaining space rocks are few, and that encounters with planets were mainly confined to the past—are being upended. On March 18, 2004, for instance, a 30-meter asteroid designated 2004 FH—a hunk potentially large enough to obliterate a city—shot past Earth, not far above the orbit occupied by telecommunications satellites. (Enter "2004 FH" in the search box at Wikipedia and you can watch film of that asteroid passing through the night sky.) Looking at the broader picture, in 1992 the astronomers David Jewitt, of the University of Hawaii, and Jane Luu, of the Massachusetts Institute of Technology, discovered the Kuiper Belt, a region of asteroids and comets that starts near the orbit of Neptune and extends for immense distances outward. At least 1,000 objects big enough to be seen from Earth have already been located there. These objects are 100 kilometers across or larger, much bigger than whatever dispatched the dinosaurs; space rocks this size are referred to as "planet killers" because their impact would likely end life on Earth. Investigation of the Kuiper Belt has just begun, but there appear to be substantially more asteroids in this region than in the asteroid belt, which may need a new name.

Beyond the Kuiper Belt may lie the hypothesized Oort Cloud, thought to contain as many as trillions of comets. If the Oort Cloud does exist, the number of extant comets is far greater than was once believed. Some astronomers now think that short-period comets, which swing past the sun frequently, hail from the relatively nearby Kuiper Belt, whereas comets whose return periods are longer originate in the Oort Cloud.

But if large numbers of comets and asteroids are still around, several billion years after the formation of the solar system, wouldn't they by now be in stable orbits—ones that rarely intersect those of the planets? Maybe not. During the past few decades, some astronomers have theorized that the movement of the solar system within the Milky Way varies the gravitational stresses to which the sun, and everything that revolves around it, is exposed. The solar system may periodically pass close to stars or groups of stars whose gravitational pull affects the Oort Cloud, shaking comets and asteroids loose from their orbital moorings and sending them downward, toward the inner planets.

Consider objects that are already near Earth, and the picture gets even bleaker. Astronomers traditionally spent little time looking for asteroids, regarding them as a lesser class of celestial bodies, lacking the beauty of comets or the significance of planets and stars. Plus, asteroids are hard to spot—they move rapidly, compared with the rest of the heavens, and even the nearby ones are fainter than other objects in space. Not until the 1980s did scientists begin systematically searching for asteroids near Earth. They have been finding them in disconcerting abundance.

In 1980, only 86 near-Earth asteroids and comets were known to exist. By 1990, the figure had risen to 170; by 2000, it was 921; as of this writing, it is 5,388. The Jet Propulsion Laboratory, part of NASA, keeps a running tally at www.neo.jpl.nasa.gov/stats. Ten years ago, 244 near-Earth space rocks one kilometer across or more—the size that would cause global calamity—were known to exist; now 741 are. Of the recently discovered nearby space objects, NASA has classified 186 as "impact risks" (details about these rocks are at www.neo.jpl.nasa.gov/risk). And because most space-rock searches to date have been low-budget affairs, conducted with equipment designed to look deep into the heavens, not at nearby space, the actual number of impact risks is undoubtedly much higher. Extrapolating from recent discoveries, NASA estimates that there are perhaps 20,000 potentially hazardous asteroids and comets in the general vicinity of Earth.

There's still more bad news. Earth has experienced several mass extinctions—the dinosaurs died about 65 million years ago, and something killed off some 96 percent of the world's marine species about 250 million years ago. Scientists have generally assumed that whatever caused those long-ago mass extinctions—comet impacts, extreme volcanic activity—arose from conditions that have changed and no longer pose much threat. It's a comforting notion—but what about the mass extinction that occurred close to our era?

About 12,000 years ago, many large animals of North America started disappearing—woolly mammoths, saber-toothed cats, mastodons, and others. Some scientists have speculated that Paleo-Indians may have hunted some of the creatures to extinction. A millennia-long mini–Ice Age also may have been a factor. But if that's the case, what explains the disappearance of the Clovis People, the best-documented Paleo-Indian culture, at about the same time? Their population stretched as far south as Mexico, so the mini–Ice Age probably was not solely responsible for their extinction.

A team of researchers led by Richard Firestone, of the Lawrence Berkeley National Laboratory, in California, recently announced the discovery of evidence that one or two huge space rocks, each perhaps several kilometers across, exploded high above Canada 12,900 years ago. The detonation, they believe, caused widespread fires and dust clouds, and disrupted climate patterns so severely that it triggered a prolonged period of global cooling. Mammoths and other species might have been killed either by the impact itself or by starvation after their food supply was disrupted. These conclusions, though hotly disputed by other researchers, were based on extensive examinations of soil samples from across the continent; in strata from that era, scientists found widely distributed soot and also magnetic grains of iridium, an element that is rare on Earth but common in space. Iridium is the meteor-hunter's lodestar: the discovery of iridium dating back 65 million years is what started the geologist Walter Alvarez on his path-breaking theory about the dinosaurs' demise.

A more recent event gives further cause for concern. As buffs of the television show *The X Files* will recall, just a century ago, in 1908, a huge explosion occurred above Tunguska, Siberia. The cause was not a malfunctioning alien star-cruiser but a small asteroid or comet that detonated as it approached the ground. The blast had hundreds of times the force of the Hiroshima bomb and devastated an area of several hundred square miles. Had the explosion occurred above London or Paris, the city would no longer exist. Mark Boslough, a researcher at the Sandia National Laboratory, in New Mexico, recently concluded that the Tunguska object was surprisingly small, perhaps only 30 meters across. Right now, astronomers are nervously tracking 99942 Apophis, an asteroid with a slight chance of striking Earth in April 2036. Apophis is also small by asteroid standards, perhaps 300 meters across, but it could hit with about 60,000 times the force of the Hiroshima bomb—enough to destroy an area the size of France. In other words, small asteroids may be more dangerous than we used to think—and may do considerable damage even if they don't reach Earth's surface.

Until recently, nearly all the thinking about the risks of space-rock strikes has focused on counting craters. But what if most impacts don't leave craters? This is the prospect that troubles Boslough. Exploding in the air, the Tunguska rock did plenty of damage, but if people had not seen the flashes, heard the detonation, and traveled to the remote area to photograph the scorched, flattened wasteland, we'd never know the Tunguska event had happened. Perhaps a comet or two exploding above Canada 12,900 years ago spelled the end for saber-toothed cats and Clovis society. But no obvious crater resulted; clues to the calamity were subtle and hard to come by.

Comets, asteroids, and the little meteors that form pleasant shooting stars approach Earth at great speeds—at least 25,000 miles per hour. As they enter the atmosphere they heat up, from friction, and compress, because they decelerate rapidly. Many space rocks explode under this stress, especially small ones; large objects are more likely to reach Earth's surface. The angle at which objects enter the atmosphere also matters: an asteroid or comet approaching straight down has a better chance of hitting the surface than one entering the atmosphere at a shallow angle, as the latter would have to plow through more air, heating up and compressing as it descended. The object or objects that may have detonated above Canada 12,900 years ago would probably have approached at a shallow angle.

If, as Boslough thinks, most asteroids and comets explode before reaching the ground, then this is another reason to fear that the conventional thinking seriously underestimates the frequency of space-rock strikes—the small number of craters may be lulling us into complacency. After all, if a space rock were hurtling toward a city, whether it would leave a crater would not be the issue—the explosion would be the issue.

A generation ago, the standard assumption was that a dangerous object would strike Earth perhaps once in a million years. By the mid-1990s, researchers began to say that the threat was greater: perhaps a strike every 300,000 years. This winter, I asked William Ailor, an asteroid specialist at The Aerospace Corporation, a think tank for the Air Force, what he thought the risk was. Ailor's answer: a 1-in-10 chance per century of a dangerous space-object strike.

Regardless of which estimate is correct, the likelihood of an event is, of course, no predictor. Even if space strikes are *likely* only once every million years, that doesn't mean a million years will pass before the next impact—the sky could suddenly darken tomorrow. Equally important, improbable but cataclysmic dangers ought to command attention because of their scope. A tornado is far more likely than an asteroid strike, but humanity is sure to survive the former. The chances that any one person will die in an airline crash are minute, but this does not prevent us from caring about aviation safety. And as Nathan Myhrvold, the former chief technology officer of Microsoft, put it, "The odds of a space-object strike during your lifetime may be no more than the odds you will die in a plane crash—but with space rocks, it's like the entire human race is riding on the plane."

Given the scientific findings, shouldn't space rocks be one of NASA's priorities? You'd think so, but Dallas Abbott says NASA has shown no interest in her group's work: "The NASA people don't want to believe me. They won't even listen."

NASA supports some astronomy to search for near-Earth objects, but the agency's efforts have been piecemeal and underfunded, backed by less than a tenth of a percent of the NASA budget. And though altering the course of space objects approaching Earth appears technically feasible, NASA possesses no hardware specifically for this purpose, has nearly nothing in development, and has resisted calls to begin work on protection against space strikes. Instead, NASA is enthusiastically preparing to spend hundreds of billions of taxpayers' dollars on a manned moon base that has little apparent justification.

"What is in the best interest of the country is never even mentioned in current NASA planning," says Russell Schweickart, one of the *Apollo* astronauts who went into space in 1969, who is leading a campaign to raise awareness of the threat posed by space rocks. "Are we going to let a space strike kill millions of people before we get serious about this?" he asks.

In January, I attended an internal NASA conference, held at agency headquarters, during which NASA's core goals were presented in a PowerPoint slideshow. Nothing was said about protecting Earth from space strikes—not even researching what sorts of spacecraft might be used in an approaching-rock emergency. Goals that *were* listed included "sustained human presence on the moon for national preeminence" and "extend the human presence across the solar system and beyond." Achieving national preeminence—isn't the United States pretty well-known already? As for extending our presence, a manned mission to Mars is at least decades away, and human travel to the outer planets is not seriously discussed by even the most zealous advocates of space exploration. Sending people "beyond" the solar system is inconceivable with any technology that can reasonably be foreseen; an interstellar spaceship traveling at the fastest speed ever achieved in space flight would take 60,000 years to reach the next-closest star system.

After the presentation, NASA's administrator, Michael Griffin, came into the room. I asked him why there had been no discussion of space rocks. He said, "We don't make up our goals. Congress has not instructed us to provide Earth defense. I administer the policy set by Congress and the White House, and that policy calls for a focus on return to the moon. Congress and the White House do not ask me what I think." I asked what NASA's priorities would be if he did set the goals. "The same. Our priorities are correct now," he answered. "We are on the right path. We need to go back to the moon. We don't need a near-Earth-objects program." In a public address about a month later, Griffin said that the moon-base plan was "the finest policy framework for United States civil space activities that I have seen in 40 years."

Actually, Congress *has* asked NASA to pay more attention to space rocks. In 2005, Congress instructed the agency to mount a sophisticated search of the proximate heavens for asteroids and comets, specifically requesting that NASA locate all near-Earth objects 140 meters or larger that are less than 1.3 astronomical units from the sun—roughly out to the orbit of Mars. Last year, NASA gave Congress its reply: an advanced search of the sort Congress was requesting would cost about $1 billion, and the agency had no intention of diverting funds from existing projects, especially the moon-base initiative.

How did the moon-base idea arise? In 2003, after the shuttle *Columbia* was lost, manned space operations were temporarily shut down, and the White House spent a year studying possible new missions for NASA. George W. Bush wanted to announce a voyage to Mars. Every Oval Office occupant since John F. Kennedy knows how warmly history has praised him for the success of his pledge to put men on the moon; it's only natural that subsequent presidents would dream about securing their own place in history by sending people to the Red Planet. But the technical barriers and even the most optimistic cost projections for a manned mission to Mars are prohibitive. So in 2004, Bush unveiled a compromise plan: a permanent moon base that would be promoted as a stepping-stone for a Mars mission at some unspecified future date. As anyone with an aerospace engineering background well knows, stopping at the moon, as Bush was suggesting, actually would be an impediment to Mars travel, because huge amounts of fuel would be wasted landing on the moon and then blasting off again. Perhaps something useful to a Mars expedition would be learned in the course of building a moon base; but if the goal is the Red Planet, then spending vast sums on lunar living would only divert that money from the research and development needed for Mars hardware. However, saying that a moon base would one day support a Mars mission allowed Bush to create the impression that his plan would not merely be restaging an effort that had already been completed more than 30 years before. For NASA, a decades-long project to build a moon base would ensure a continuing flow of money to its favorite contractors and to the congressional districts where manned-space-program centers are located. So NASA signed on to the proposal, which Congress approved the following year.

It is instructive, in this context, to consider the agency's rhetoric about China. The Chinese manned space program has been improving and is now about where the U.S. program was in the mid-1960s. Stung by criticism that the moon-base project has no real justification—37 years ago, President Richard Nixon cancelled the final planned *Apollo* moon missions because the program was accomplishing little at great expense; as early as 1964, the communitarian theorist Amitai Etzioni was calling lunar obsession a "moondoggle"—NASA is selling the new plan as a second moon race, this time against Beijing. "I'll be surprised if the Chinese don't reach the moon before we return," Griffin said. "China is now a strategic peer competitor to the United States in space. China is drawing national prestige from achievements in space, and there will be a tremendous shift in national prestige toward Beijing if the Chinese are operating on the moon and we are not. Great nations have always operated on the frontiers of their era. The moon is the frontier of our era, and we must outperform the Chinese there."

Wouldn't shifting NASA's focus away from wasting money on the moon and toward something of clear benefit for the entire world—identifying and deflecting dangerous space objects—be a surer route to enhancing national prestige? But NASA's institutional instinct is not to ask, "What can we do in space that makes sense?" Rather, it is to ask, "What can we do in space that requires lots of astronauts?" That finding and stopping space rocks would be an expensive mission with little role for the astronaut corps is, in all likelihood, the principal reason NASA doesn't want to talk about the asteroid threat.

NASA's lack of interest in defending against space objects leaves a void the Air Force seems eager to fill. The Air Force has the world's second-largest space program, with a budget of about $11 billion—$6 billion less than NASA's. The tension between the two entities is long-standing. Many in the Air Force believe the service could achieve U.S. space objectives faster and more effectively than NASA. And the Air Force simply wants flyboys in orbit: several times in the past, it has asked Congress to fund its own space station, its own space plane, and its own space-shuttle program. Now, with NASA all but ignoring the space-object threat, the Air Force appears to be seizing an opportunity.

All known space rocks have been discovered using telescopes designed for traditional "soda straw" astronomy—that is, focusing on a small patch of sky. Now the Air Force is funding the first research installation designed to conduct panoramic scans of the sky, a telescope complex called Pan-STARRS, being built by the University of Hawaii. By continuously panning the entire sky, Pan-STARRS should be able to spot many near-Earth objects that so far have gone undetected. The telescope also will have substantially better resolving power and sensitivity than existing survey instruments, enabling it to find small space rocks that have gone undetected because of their faintness.

The Pan-STARRS project has no military utility, so why is the Air Force the sponsor? One speculation is that Pan-STARRS is the Air Force's foot in the door for the Earth-defense mission. If the Air Force won funding to build high-tech devices to fire at asteroids, this would be a major milestone in its goal of an expanded space presence. But space rocks are a natural hazard, not a military threat, and an Air Force Earth-protection initiative, however gallant, would probably cause intense international opposition. Imagine how other governments would react if the Pentagon announced, "Don't worry about those explosions in space—we're protecting you."

Thus, the task of defending Earth from objects falling from the skies seems most fitting for NASA, or perhaps for a multinational civilian agency that might be created. Which raises the question: What could NASA, or anyone else, actually do to provide a defense?

RUSSELL SCHWEICKART, THE FORMER *Apollo* astronaut, runs the B612 Foundation (B612 is the asteroid home of Saint-

Exupéry's Little Prince). The foundation's goal is to get NASA officials, Congress, and ultimately the international community to take the space-rock threat seriously; it advocates testing a means of precise asteroid tracking, then trying to change the course of a near-Earth object.

Current telescopes cannot track asteroids or comets accurately enough for researchers to be sure of their courses. When 99942 Apophis was spotted, for example, some calculations suggested it would strike Earth in April 2029, but further study indicates it won't—instead, Apophis should pass between Earth and the moon, during which time it may be visible to the naked eye. The Pan-STARRS telescope complex will greatly improve astronomers' ability to find and track space rocks, and it may be joined by the Large Synoptic Survey Telescope, which would similarly scan the entire sky. Earlier this year, the software billionaires Bill Gates and Charles Simonyi pledged $30 million for work on the LSST, which proponents hope to erect in the mountains of Chile. If it is built, it will be the first major telescope to broadcast its data live over the Web, allowing countless professional and amateur astronomers to look for undiscovered asteroids.

Schweickart thinks, however, that even these instruments will not be able to plot the courses of space rocks with absolute precision. NASA has said that an infrared telescope launched into an orbit near Venus could provide detailed information on the exact courses of space rocks. Such a telescope would look outward from the inner solar system toward Earth, detect the slight warmth of asteroids and comets against the cold background of the cosmos, and track their movements with precision. Congress would need to fund a near-Venus telescope, though, and NASA would need to build it—neither of which is happening.

Another means of gathering data about a potentially threatening near-Earth object would be to launch a space probe toward it and attach a transponder, similar to the transponders used by civilian airliners to report their exact locations and speed; this could give researchers extremely precise information on the object's course.

There is no doubt that a probe can rendezvous with a space rock: in 2005, NASA smashed a probe called *Deep Impact* into the nucleus of comet 9P/Tempel in order to vaporize some of the material on the comet's surface and make a detailed analysis of it. Schweickart estimates that a mission to attach a transponder to an impact-risk asteroid could be staged for about $400 million—far less than the $11.7 billion cost to NASA of the 2003 *Columbia* disaster.

Then what? In the movies, nuclear bombs are used to destroy space rocks. In NASA's 2007 report to Congress, the agency suggested a similar approach. But nukes are a brute-force solution, and because an international treaty bans nuclear warheads in space, any proposal to use them against an asteroid would require complex diplomatic agreements. Fortunately, it's likely that just causing a slight change in course would avert a strike. The reason is the mechanics of orbits. Many people think of a planet as a vacuum cleaner whose gravity sucks in everything in its vicinity. It's true that a free-falling body will plummet toward the nearest source of gravity—but in space, free-falling bodies are rare. Earth does not plummet into the sun, because the angular momentum of Earth's orbit is in equilibrium with the sun's gravity. And asteroids and comets swirl around the sun with tremendous angular momentum, which prevents them from falling toward most of the bodies they pass, including Earth.

For any space object approaching a planet, there exists a "keyhole"—a patch in space where the planet's gravity and the object's momentum align, causing the asteroid or comet to hurtle toward the planet. Researchers have calculated the keyholes for a few space objects and found that they are tiny, only a few hundred meters across—pinpoints in the immensity of the solar system. You might think of a keyhole as the win-a-free-game opening on the 18th tee of a cheesy, incredibly elaborate miniature-golf course. All around the opening are rotating windmills, giants stomping their feet, dragons walking past, and other obstacles. If your golf ball hits the opening precisely, it will roll down a pipe for a hole in one. Miss by even a bit, and the ball caroms away.

Tiny alterations might be enough to deflect a space rock headed toward a keyhole. "The reason I am optimistic about stopping near-Earth-object impacts is that it looks like we won't need to use fantastic levels of force," Schweickart says. He envisions a "gravitational tractor," a spacecraft weighing only a few tons—enough to have a slight gravitational field. If an asteroid's movements were precisely understood, placing a gravitational tractor in exactly the right place should, ever so slowly, alter the rock's course, because low levels of gravity from the tractor would tug at the asteroid. The rock's course would change only by a minuscule amount, but it would miss the hole-in-one pipe to Earth.

Will the gravitational-tractor idea work? The B612 Foundation recommends testing the technology on an asteroid that has no chance of approaching Earth. If the gravitational tractor should prove impractical or ineffective, other solutions could be considered. Attaching a rocket motor to the side of an asteroid might change its course. So might firing a laser: as materials boiled off the asteroid, the expanding gases would serve as a natural jet engine, pushing it in the opposite direction.

But when it comes to killer comets, you'll just have to lose sleep over the possibility of their approach; there are no proposals for what to do about them. Comets are easy to see when they are near the sun and glowing but are difficult to detect at other times. Many have "eccentric" orbits, spending centuries at tremendous distances from the sun, then falling toward the inner solar system, then slingshotting away again. If you were to add comets to one of those classroom models of the solar system, many would need to come from other floors of the building, or from another school district, in order to be to scale.

Advanced telescopes will probably do a good job of detecting most asteroids that pass near Earth, but an unknown comet suddenly headed our way would be a nasty surprise. And because many comets change course when the sun heats their sides and causes their frozen gases to expand, deflecting or destroying them poses technical problems to which there are no ready solutions. The logical first step, then, seems to be to determine how to prevent an asteroid from striking Earth and hope that some future advance, perhaps one building on the asteroid work, proves useful against comets.

None of this will be easy, of course. Unlike in the movies, where impossibly good-looking, wisecracking men and women grab space suits and race to the launchpad immediately after receiving a warning that something is approaching from space, in real life preparations to defend against a space object would take many years. First the necessary hardware must be built—quite possibly a range of space probes and rockets. An asteroid that appeared to pose a serious risk would require extensive study, and a transponder mission could take years to reach it. International debate and consensus would be needed: the possibility of one nation acting alone against a space threat or of, say, competing U.S. and Chinese missions to the same object, is more than a little worrisome. And suppose Asteroid X appeared to threaten Earth. A mission by, say, the United States to deflect or destroy it might fail, or even backfire, by nudging the rock toward a gravitational keyhole rather than away from it. Asteroid X then hits Costa Rica; is the U.S. to blame? In all likelihood, researchers will be unable to estimate where on Earth a space rock will hit. Effectively, then, everyone would be threatened, another reason nations would need to act cooperatively—and achieving international cooperation could be a greater impediment than designing the technology.

WE WILL SOON HAVE A new president, and thus an opportunity to reassess NASA's priorities. Whoever takes office will decide whether the nation commits to spending hundreds of billions of dollars on a motel on the moon, or invests in space projects of tangible benefit—space science, environmental studies of Earth, and readying the world for protection against a space-object strike. Although the moon-base initiative has been NASA's focus for four years, almost nothing has yet been built for the project, and comparatively little money has been spent; current plans don't call for substantial funding until the space-shuttle program ends, in 2010. This suggests that NASA could back off from the moon base without having wasted many resources. Further, the new Ares rocket NASA is designing for moon missions might be just the ticket for an asteroid-deflection initiative.

Congress, too, ought to look more sensibly at space priorities. Because it controls federal funding, Congress holds the trump cards. In 2005, it passively approved the moon-base idea, seemingly just as budgetary log-rolling to maintain spending in the congressional districts favored under NASA's current budget hierarchy. The House and Senate ought to demand that the space program have as its first priority returning benefits to taxpayers. It's hard to imagine how taxpayers could benefit from a moon base. It's easy to imagine them benefiting from an effort to protect our world from the ultimate calamity.

DENNIS OVERBYE

Big Brain Theory: Have Cosmologists Lost Theirs?

FROM *THE NEW YORK TIMES*

What happens when cosmologists try to think through the staggering implications of a truly infinite universe? Disembodied brains float through space—or, at least, theory says they could. Dennis Overbye tracks the real brains who are trying to solve the latest mind-bender in physics.

IT COULD BE THE WEIRDEST AND MOST EMBARRASSING prediction in the history of cosmology, if not science.

If true, it would mean that you yourself reading this article are more likely to be some momentary fluctuation in a field of matter and energy out in space than a person with a real past born through billions of years of evolution in an orderly star-spangled cosmos.

Your memories and the world you think you see around you are illusions.

This bizarre picture is the outcome of a recent series of calculations that take some of the bedrock theories and discoveries of modern cosmology to the limit. Nobody in the field believes that this is the way things really work, however. And so in the last couple of years there has been a growing stream of debate and dueling papers, replete with references to such esoteric subjects as reincarnation, multiple universes and even the death of spacetime, as cosmologists try to square the predictions of their cherished theories with their convictions that we and the universe are real. The basic problem is that across the eons of time, the standard theories suggest, the universe can recur over and over again in an endless cycle of big bangs, but it's hard for nature to make a whole universe. It's much easier to make fragments of one, like planets, yourself maybe in a spacesuit or even—in the most absurd and troubling example—a naked brain floating in space. Nature tends to do what is easiest, from the standpoint of energy and probability. And so these fragments—in particular the brains—would appear far more frequently than real full-fledged universes, or than us. Or they might *be* us.

Alan Guth, a cosmologist at the Massachusetts Institute of Technology who agrees this overabundance is absurd, pointed out that some calculations result in an infinite number of free-floating brains for every normal brain, making it "infinitely unlikely for us to be normal brains." Welcome to what physicists call the Boltzmann brain problem, named after the 19th-century Austrian physicist Ludwig Boltzmann, who suggested the mechanism by which such fluctuations could happen in a gas or in the universe. Cosmologists also refer to them as "freaky observers," in contrast to regular or "ordered" observers of the cosmos like ourselves. Cosmologists are desperate to eliminate these freaks from their theories, but so far they can't even agree on how or even on whether they are making any progress.

If you are inclined to skepticism this debate might seem like further evidence that cosmologists, who gave us dark matter, dark energy and speak with apparent aplomb about gazillions of parallel universes, have finally lost their minds. But the cosmologists say the brain problem serves as a valuable reality check as they contemplate the far, far future and zillions of bubble universes popping off from one another in an ever-increasing rush through eternity. What, for example, is a "typical" observer in such a setup? If some atoms in another universe stick together briefly to look, talk and think exactly like you, is it really you?

"It is part of a much bigger set of questions about how to think about probabilities in an infinite universe in which everything that can occur, does occur, infinitely many times," said Leonard Susskind of Stanford, a co-author of a paper in 2002 that helped set off the debate. Or as Andrei Linde, another Stanford theorist given to colorful language, loosely characterized the possibility of a replica of your own brain forming out in space sometime, "How do you compute the probability to be reincarnated to the probability of being born?"

The Boltzmann brain problem arises from a string of logical conclusions that all spring from another deep and old question, namely why time seems to go in only one direction. Why can't you unscramble an egg? The fundamental laws governing the atoms bouncing off one another in the egg look the same whether time goes forward or backward. In this universe, at least, the future and the past are different and you can't remember who is going to win the Super Bowl next week.

"When you break an egg and scramble it you are doing cosmology," said Sean Carroll, a cosmologist at the California Institute of Technology.

Boltzmann ascribed this so-called arrow of time to the tendency of any collection of particles to spread out into the most random and useless configuration, in accordance with the second law of thermodynamics (sometimes paraphrased as "things get worse"), which says that entropy, which is a measure of disorder or wasted energy, can never decrease in a closed system like the universe.

If the universe was running down and entropy was increasing

now, that was because the universe must have been highly ordered in the past.

In Boltzmann's time the universe was presumed to have been around forever, in which case it would long ago have stabilized at a lukewarm temperature and died a "heat death." It would already have maximum entropy, and so with no way to become more disorderly there would be no arrow of time. No life would be possible but that would be all right because life would be excruciatingly boring. Boltzmann said that entropy was all about odds, however, and if we waited long enough the random bumping of atoms would occasionally produce the cosmic equivalent of an egg unscrambling. A rare fluctuation would decrease the entropy in some place and start the arrow of time pointing and history flowing again. That is not what happened. Astronomers now know the universe has not lasted forever. It was born in the Big Bang, which somehow set the arrow of time, 14 billion years ago. The linchpin of the Big Bang is thought to be an explosive moment known as inflation, during which space became suffused with energy that had an antigravitational effect and ballooned violently outward, ironing the kinks and irregularities out of what is now the observable universe and endowing primordial chaos with order.

Inflation is a veritable cosmological fertility principle. Fluctuations in the field driving inflation also would have seeded the universe with the lumps that eventually grew to be galaxies, stars and people. According to the more extended version, called eternal inflation, an endless array of bubble or "pocket" universes are branching off from one another at a dizzying and exponentially increasing rate. They could have different properties and perhaps even different laws of physics, so the story goes.

A different, but perhaps related, form of antigravity, glibly dubbed dark energy, seems to be running the universe now, and that is the culprit responsible for the Boltzmann brains.

The expansion of the universe seems to be accelerating, making galaxies fly away from one another faster and faster. If the leading dark-energy suspect, a universal repulsion Einstein called the cosmological constant, is true, this runaway process will last forever, and distant galaxies will eventually be moving apart so quickly that they cannot communicate with one another. Being in such a space would be like being surrounded by a black hole.

Rather than simply going to black like *The Sopranos* conclusion, however, the cosmic horizon would glow, emitting a feeble spray of elementary particles and radiation, with a temperature of a fraction of a billionth of a degree, courtesy of quantum uncertainty. That radiation bath will be subject to random fluctuations just like Boltzmann's eternal universe, however, and every once in a very long, long time, one of those fluctuations would be big enough to recreate the Big Bang. In the fullness of time this process could lead to the endless series of recurring universes. Our present universe could be part of that chain.

In such a recurrent setup, however, Dr. Susskind of Stanford, Lisa Dyson, now of the University of California, Berkeley, and Matthew Kleban, now at New York University, pointed out in 2002 that Boltzmann's idea might work too well, filling the megaverse with more Boltzmann brains than universes or real people.

In the same way the odds of a real word showing up when you shake a box of Scrabble letters are greater than a whole sentence or paragraph forming, these "regular" universes would be vastly outnumbered by weird ones, including flawed variations on our own all the way down to naked brains, a result foreshadowed by Martin Rees, a cosmologist at the University of Cambridge, in his 1997 book, *Before the Beginning*.

The conclusions of Dr. Dyson and her colleagues were quickly challenged by Andreas Albrecht and Lorenzo Sorbo of the University of California, Davis, who used an alternate approach. They found that the Big Bang was actually more likely than Boltzmann's brain.

"In the end, inflation saves us from Boltzmann's brain," Dr. Albrecht said, while admitting that the calculations were contentious. Indeed, the "invasion of Boltzmann brains," as Dr. Linde once referred to it, was just beginning.

In an interview Dr. Linde described these brains as a form of reincarnation. Over the course of eternity, he said, anything is possible. After some Big Bang in the far future, he said, "it's possible that you yourself will re-emerge. Eventually you will appear with your table and your computer."

But it's more likely, he went on, that you will be reincarnated as an isolated brain, without the baggage of stars and galaxies. In terms of probability, he said, "It's cheaper."

YOU MIGHT WONDER WHAT'S WRONG with a few brains—or even a preponderance of them—floating around in space. For one thing, as observers these brains would see a freaky chaotic universe, unlike our own, which seems to persist in its promise and disappointment.

Another is that one of the central orthodoxies of cosmology is that humans don't occupy a special place in the cosmos, that we and our experiences are typical of cosmic beings. If the odds of us being real instead of Boltzmann brains are one in a million, say, waking up every day would be like walking out on the street and finding everyone in the city standing on their heads. You would expect there to be some reason why you were the only one left right side up.

Some cosmologists, James Hartle and Mark Srednicki, of the University of California, Santa Barbara, have questioned that assumption. "For example," Dr. Hartle wrote in an e-mail message, "on Earth humans are not typical animals; insects are far more numerous. No one is surprised by this."

In an e-mail response to Dr. Hartle's view, Don Page of the University of Alberta, who has been a prominent voice in the Boltzmann debate, argued that what counted cosmologically was not sheer numbers, but consciousness, which we have in abundance over the insects. "I would say that we have no strong evidence against the working hypothesis that we are typical and that our observations are typical," he explained, "which is very fruitful in science for helping us believe that our observations are not just flukes but do tell us something about the universe."

Dr. Dyson and her colleagues suggested that the solution to the Boltzmann paradox was in denying the presumption that the universe would accelerate eternally. In other words, they said, that the cosmological constant was perhaps not really constant. If the cosmological constant eventually faded away, the universe would revert to normal expansion and what was left would eventually fade to black. With no more acceleration there would be no horizon with its snap, crackle and pop, and thus no material for fluctuations and Boltzmann brains.

String theory calculations have suggested that dark energy is indeed metastable and will decay, Dr. Susskind pointed out. "The success of ordinary cosmology," Dr. Susskind said, "speaks against the idea that the universe was created in a random fluctuation."

But nobody knows whether dark energy—if it dies—will die soon enough to save the universe from a surplus of Boltzmann brains. In 2006, Dr. Page calculated that the dark energy would have to decay in about 20 billion years in order to prevent it from being overrun by Boltzmann brains.

The decay, if and when it comes, would rejigger the laws of physics and so would be fatal and total, spreading at almost the speed of light and destroying all matter without warning. There would be no time for pain, Dr. Page wrote: "And no grieving survivors will be left behind. So in this way it would be the most humanely possible execution." But the object of his work, he said, was not to predict the end of the universe but to draw attention to the fact that the Boltzmann brain problem remains.

People have their own favorite measures of probability in the multiverse, said Raphael Bousso of the University of California, Berkeley. "So Boltzmann brains are just one example of how measures can predict nonsense; anytime your measure predicts that something we see has extremely small probability, you can throw it out," he wrote in an e-mail message.

Another contentious issue is whether the cosmologists in their calculations could consider only the observable universe, which is all we can ever see or be influenced by, or whether they should take into account the vast and ever-growing assemblage of other bubbles forever out of our view predicted by eternal inflation. In the latter case, as Alex Vilenkin of Tufts University pointed out, "The numbers of regular and freak observers are both infinite." Which kind predominate depends on how you do the counting, he said.

In eternal inflation, the number of new bubbles being hatched at any given moment is always growing, Dr. Linde said, explaining one such counting scheme he likes. So the evolution of people in new bubbles far outstrips the creation of Boltzmann brains in old ones. The main way life emerges, he said, is not by reincarnation but by the creation of new parts of the universe. "So maybe we don't need to care too much" about the Boltzmann brains, he said.

"If you are reincarnated, why do you care about where you are reincarnated?" he asked. "It sounds crazy because here we are touching issues we are not supposed to be touching in ordinary science. Can we be reincarnated?"

"People are not prepared for this discussion," Dr. Linde said.

KAREN OLSSON

The Final Frontier

FROM *THE TEXAS MONTHLY*

Two astronomers based in Texas want to study the secrets of the universe. But it costs money. To raise that money, they look beyond the federal-grant system, to wealthy individuals and others. Karen Olsson watches as two scientists become salesmen—and in Texas, no less.

ONE EVENING LAST OCTOBER, A UNIVERSITY OF TEXAS at Austin astronomer named Gary Hill stood behind a lectern at Miss Hattie's Café and Saloon, which occupies a restored nineteenth-century bank building in downtown San Angelo, and cheerfully proclaimed his ignorance. "We really don't understand the universe," he said. "We thought we did, but it turns out we only understand about four percent of it."

A dozen or so people, among them businessmen, the editor of the local newspaper, a school librarian, and a couple of college professors, had assembled at Miss Hattie's, where the lace curtains and rose-print wallpaper harked back to a time when the universe was no larger than our own galaxy and Newton's laws seemed to explain it and a tunnel linked the building where we sat to a nearby bordello. There was something old-fashioned too about the fact that a bespectacled, British-born scientist had traveled from the state capital to give a talk to the curious and that the curious had turned out to hear what he had to say. His subject, on the other hand, was the very future of cosmology and physics and how they might be affected by one telescope in particular, located 212 miles farther west.

Galileo stuck lenses onto either end of an organ pipe; today's research telescopes, while considerably more elaborate, still perform the same fundamental task of collecting and focusing light. It's all astronomers have to go on: electromagnetic radiation from distant objects, whether it arrives in the form of X-rays or visible light or radio waves. "We're detectives, but we can only use what light will give us," Hill had said to me earlier that day. "So we get fairly ingenious in the ways we analyze light to look for clues." They rely, for instance, on spectroscopy, the process of separating light emitted by an object in space into its component wavelengths, as a prism does, then analyzing those components. And they invent new tools to analyze the light. To probe deeper and deeper into space, scientists must design better and better detectors, sensitive to the faintest of emissions.

Such instruments don't come cheaply, which is why Hill and his colleague Karl Gebhardt have periodically taken to the road over the past three and a half years: They've been promoting an ambitious $34 million overhaul of a telescope at UT's McDonald Observatory, in the Davis Mountains of West Texas. Speaking to potential donors in Houston or a luncheon group in Abilene, they've been publicizing an endeavor called HETDEX, or the Hobby-Eberly Telescope Dark Energy Experiment, the aim of which is to help attack what some have labeled one of the most important problems in science.

Hill grew up in England but left to go to where the telescopes were, first as a graduate student in Hawaii, then as a postdoctoral researcher in West Texas, in 1988. Finding that certain instruments at the observatory weren't sensitive enough, he designed and built a new spectrograph on the cheap, still in use today. He is now the observatory's chief astronomer. At Miss Hattie's he applied his knack for innovative thinking to the problem of business attire—he wore a striped green shirt and a pink tie—and as he spoke, he grinned and nodded infectiously. "We have a huge opportunity to lay the groundwork in Texas for understanding how the universe has changed through time," he said. He outlined the goals of the experiment: to conduct the largest survey of other galaxies ever completed and to use that information to measure how the scale of the universe has evolved—and to reinvent the telescope in the process. By doing so, Hill, Gebhardt, and their collaborators hope to better understand what astronomers call dark energy, though no one really knows what the term means: "Dark energy" is a label for a mystery. "The thing is," Hill told his audience, "it may not be dark and it may not be energy."

After the presentation had ended and most of the audience had departed, Hill was subjected to a more rigorous interrogation at dinner from Ken Gunter, a tall, poker-faced San Angelo businessman and a member of the McDonald's Board of Visitors, a statewide group of observatory supporters. This turned into a debate between the gruff West Texan and the polite but impassioned British scientist, while half a dozen others at the table looked on. Though he supports the experiment, Gunter was skeptical as to whether he could raise funds for it. "What is the pragmatic end product, except exciting a bunch of astronomy Ph.D.'s in Austin? Give me something I can identify with. If you want to raise some honest-to-God money, you better start raising some honest-to-God tie back to medicine or energy." That, said Hill, wasn't the point. The most practical argument he could make for the telescope was that it might excite students in a country where science education was failing and draw better faculty to the university. Gunter seemed unconvinced. "My sense is that most people don't give two hoots in hell whether the universe is expanding or contracting or moving sideways!" he said, and drew out a cigar.

Earlier that day I'd ridden in a rented suburban with Hill; David Lambert, the director of the observatory; and Joel Barna, its development director, from Midland to Abilene to San Angelo. The flat landscape was staked by the technology of energy production: Near Midland the pump jacks kowtowed to the brush, while farther south a line of soaring silver windmills receded toward the horizon. (One question often posed by lay audiences in Texas, Hill told me, is "How can we harness dark energy for human use?") It was a warm, hazy morning, and as we'd all risen early to catch a seven o'clock flight from Austin to Midland, a soporific air had fallen over the car, leaving me in a state of sleepy wonderment at one of astronomy's fundamental principles, that light can ripple for billions of years through the vast universe and collide with nothing else during that unfathomably long journey, reaching our planet with information about its place of origin. How can that be, I asked. "There's a lot of empty space out there," Hill said. "The fraction of the area of the sky with stuff is not very great, so the chances of a photon of light hitting anything along its route other than your telescope are actually very small."

I fell silent. All that void. A little while later I asked something else, then mulled over the answer, and that's pretty much how the ride went, as I surfaced every so often with another question for the patient professors, then let my head spin for a while. It's humbling to stumble up against the edifice of astronomy, the massive and intricate body of knowledge humans have built up over the millennia, and then all the more humbling—if somehow also reassuring—to contemplate how much more we don't know. And how much we can't possibly see, since only 4 percent of the stuff in the universe is thought to be visible matter. That includes you and me and microorganisms and other galaxies, at least 100 billion of them. As one scientist put it recently, "we're just a bit of pollution," while most of the universe is made of something else.

"Nature and nature's laws lay hid in night" goes the famous epitaph Alexander Pope composed for Isaac Newton. "God said, 'Let Newton be!' and all was light." But as it turns out, almost all is, in fact, dark, and a crucial portion of those laws remain hidden. Theorists first ventured into the "dark" to resolve a problem concerning the motions of galaxies and the stars within them. The trouble was, there just weren't enough visible stars out there. Matter far away from the center of a galaxy tended to move much faster than could be explained by the net gravitational attraction of all the visible matter. So in order to save Newton's laws of gravity, astronomers invented a new type of matter that was dark but plentiful: dark matter. The universe is littered with it, they concluded. Which might seem like cheating—the invisible check is in the mail—yet the theory has helped explain the motions of other cosmic entities, and there are high hopes that dark matter will be detected in experiments this year at CERN, the particle physics laboratory in Geneva, Switzerland.

Then scientists discovered another anomaly, this one not limited to objects like galaxies but pertaining to the entire cosmos. And they came up with something even stranger to explain it. The problem was the way in which the universe was growing. Until the twentieth century, the universe was generally thought to be static, neither expanding nor contracting. But along came Edwin Hubble (who, late bloomers take note, taught high school and coached basketball in Indiana before returning to school for his Ph.D.). In the twenties he made measurements of how far away galaxies were and how fast they were moving, and he discovered that the farther away they were, the faster they were receding. The universe, this implied, was expanding, everything moving away from everything else. (Albeit with negligible impact locally. In *Annie Hall*, Alvy Singer's mother is basically correct when she tells her existentially morose son that "Brooklyn is not expanding.") The concept of an expanding universe in turn suggested that at one time, everything was closer together. Thus Hubble's discoveries were a prod to the subsequent development of the big bang theory.

Meanwhile, a decade before Hubble, Einstein had revised Newton's idea of gravity—an attractive force between massive objects—with his theory of general relativity, in which gravity depends on the curvature of four-dimensional space-time. The twentieth century saw the Newtonian model of a fixed, orderly universe supplanted by that of a warping, expanding space with unimaginably chaotic origins. This universe had a beginning and would perhaps come to an end as well: If the expansion of space was indeed due to the propulsion from the big bang, then it stood to reason that because the gravitational pull of all the matter in the universe would be retarding that outward growth, one of three things would ultimately happen: Gravity would win out and the universe would collapse, the initial propulsion from the big bang would win and the universe would expand forever, or—and this is the most accepted case—the two would be perfectly balanced.

But in 1998, two groups of researchers independently arrived at a surprising result, so surprising that members of each originally believed that they had made a mistake. Both teams had been studying a class of exploding stars called type 1a supernovae, measuring their distances and speeds. What each team found was that the most distant supernovae were fainter than expected, fainter than they would have appeared in a decelerating or even a coasting universe. The expansion of the universe, this meant, seemed not to be slowing down but rather speeding up. And no one had a good explanation for it. What phenomenon could possibly be pushing space outward?

"Theorists have been guiding observers for ten thousand years," said Edward "Rocky" Kolb, a professor of astronomy and astrophysics at the University of Chicago. "Now they've observed something that we don't understand. We desperately need a better idea of what's going on." The notion that some form of mass energy might pervade all of space dates back to Einstein, but it gained widespread acceptance only in response to the observed acceleration of the universe. But what sort of energy? Has it changed over time? Or could it be that there really is no such thing as dark energy and that our understanding of gravity is actually wrong? These questions demanded new experiments, which in turn would require new telescopes or modifications to existing ones.

KARL GEBHARDT ARRIVED AT THE University of Texas in 2000, an expert on nearby galaxies and black holes. But in 2004 he attended a meeting on the future of U.S. telescopes, where much of the talk was about dark energy.

"It really hit home: Everyone is pushing on this," he told me when I visited him in his office. "All the dark energy missions were beginning to take root, and I said, 'Hey, look, I think we can do this at Texas.' I came back from the meeting all excited. So one day in the hall I was talking to Gary Hill about it, and I said, 'Can we do this? Can we study dark energy?' It turned out he was working on an instrument design with Phillip [MacQueen, the observatory's chief scientist and a senior researcher at UT-Austin], and we decided we can make a really nice instrument to look at the problem."

For the next few weeks, Gebhardt and Hill traded ideas about how to proceed, and MacQueen and Hill consulted about whether the instrument they'd conceived would be up to the task. "We went back and forth for a few months until we came up with a nice design," Gebhardt said. "But it's easy to come up with designs. The hard part is finding the money and the telescope to do it on." Here, though, the team had two advantages: the Hobby-Eberly Telescope and a recent grant from the Air Force.

Not long afterward, they began trying to raise the $34 million, more than the cost of the original telescope. The financial end was both a burden and a potential edge: The Texas scientists would not be bound by the same sort of bureaucratic limitations that they might have encountered using a federally administered facility or depending heavily on government grants. "The National Science Foundation won't go off the beaten path until they've beaten the path," said Gary Bernstein, a professor of physics and astronomy at the University of Pennsylvania. "If you have your own facility and can do what you want to do, you have more freedom." Or as Hill said to me, "In Texas, if you have the idea and you have the money, you can do it." And the $34 million, while hardly a small sum, was much lower than the costs of other proposed dark-energy experiments.

A plan to put together a world-class instrument on the cheap might sound an alarm in the mind of anyone familiar with the observatory's history, for a cost-saving design had been trumpeted once before, in building the Hobby-Eberly Telescope itself. Though less expensive than other telescopes of comparable size, the HET had suffered technical problems for several years after it was dedicated, in 1997. While it was one of the largest telescopes in the world, it had gained a reputation for poor image quality. Now, at last, it was functioning properly, and here came some upstart astronomers with a proposal to chop off its top half and replace the detector with a novel apparatus. Could grand ambitions get the better of the observatory a second time?

If the scientists themselves had such doubts, they kept them quiet. Their concerns were more concrete: Raise the money and get results before anyone else. Among the other proposed dark-energy experiments, only one will employ the same technique as HETDEX, and that's a collaboration between scientists in Japan and the United States to build something called the Wide-Field Multi-Object Spectrograph, which would be installed on the Japanese-operated Subaru Telescope, in Hawaii. "This is a strange game, in that you don't know exactly how much data you have to take before you can make a great discovery," Kolb said. "Timing is of the essence. It's sort of a race between the Japanese and Texans. This is such a hot topic that many people are interested in it. There is the potential of great discoveries to be made, and so you just can't sit on it and wait."

The HETDEX researchers, led by the Texas group and further

supported by two universities in Germany and the consortium that operates the HET, are the mavericks of the dark-energy industry, relying on a small number of people and a relatively low-cost instrument. So far they've raised $20.1 million—enough to build a prototype for the elaborate new spectrograph and to rebuild the HET to give it a much wider field of view—and they're gunning to finish their upgrade and their observations by 2013. "We are dwarfed by these other teams," Gebhardt said. "They have the ability to go out there and barrage everybody at conferences. We can't do that. We are working all-out just to get our instrument built. But it never hurts to be the underdog. I think people are a little scared of HETDEX because we're going to finish soon now. We are really moving."

IF YOU WERE TO LOOK at a nighttime satellite image of Texas, you would see large splotches of light produced by Dallas–Fort Worth and Houston, smaller ones for Austin and San Antonio, and a grid of glowing points as you move west from Austin, the most prominent strand of them following Interstate 10. Gradually the lights trail away; in West Texas south of the interstate, the map goes almost completely dark. It's one of the darkest spots in the continental United States, making it a natural place for an observatory.

The location was identified in the summer of 1932 by two astronomers who drove a Chevrolet nearly eight thousand miles—roughly one third of the earth's circumference—all within Texas. They ranged from Galveston to El Paso and as far north as Amarillo, stopping to peer through a small telescope, make notes, and take photographs at prospective sites. An East Texas banker named William McDonald had bequeathed his fortune to the University of Texas so that it could build an astronomical observatory (for the stated purpose of "seeing closer the gates of heaven"), and the university, which had no astronomy department, contracted with scientists from the University of Chicago to supply the expertise—the first order of business being to choose a location. They selected the

Davis Mountains, a choice approved by Otto Struve, the Russian émigré who became the observatory's first director, after he spent several nights camping on Mount Locke with his wife.

Today, as you drive from the interstate to the town of Fort Davis and on to the observatory, the elevation rises, the land begins to buckle and swell, and having passed over miles of desert plains, you find yourself driving through cedar-sprinkled hills and finally up Mount Locke, where most of the observatory is located. At first you see two white domes resembling centurion helmets at the summit. They are the observatory's older telescopes, the Struve telescope, or "the 82-inch," after the diameter of its primary mirror, which was built in 1939, and the Harlan J. Smith 107-inch, completed in 1968. Each in its time was one of the most powerful telescopes in the world, only to be eclipsed, inevitably, by technology's advance. (They are still in use, but they are no longer considered cutting-edge.) Higher up the hill a silver-mirrored dome resembling a giant buckyball comes into view: This, situated on an adjacent peak, is the Hobby-Eberly Telescope.

With its own water supply, fire marshal, staff homes, and lodge for visitors, formerly called the Transients' Quarters (now the Astronomers' Lodge), the observatory is its own little campus on a hill, one that abides by strange customs. People drive around at night with only their parking lights on, for instance, to keep things dark. It's a quiet place, where an influential portion of the population sleeps during the day and works at night, and the sky commands attention—not just from astronomers but from anyone who finds herself on the mountain. The night I arrived at the observatory, a woman passing me on the road told me to be sure to take in the stunning sunset; the next night another person bid me look out at the storms to the west.

But there's the sky we see, and then there's the sky astronomers investigate. Enter through a side door of the HET building, and it's as if you've stepped into a small manufacturing plant, with a concrete loading area and a flight of metal stairs leading to the site manager's spare office. The manager in this case is Bob Calder, who worked for the Subaru telescope, in Hawaii, and the Smithsonian Astrophysical Observatory, in Cambridge, Massachusetts, before moving to Texas to oversee the operation and maintenance of the HET—though "maintenance" is perhaps too plain a word, too reminiscent of boilers or car engines, to convey the kind of care one has to take with a major telescope. Even cleaning is a delicate task. To focus on the very thin slices of sky required for astronomical work, the telescope's large primary mirror must be absolutely smooth, free of any imperfections larger than one tenth the wavelength of the light hitting it. The aluminum coatings on the HET's mirrors are the thickness of oil on water; they are cleaned on Mondays, Wednesdays, and Fridays by a four-person team that applies a carbon dioxide "snow" with a wand.

As site manager of the HET, Calder also greets the dignitaries and potential donors who come to tour the telescope, a duty that has, on occasion, served to remind him that he is not in Cambridge any longer. "We had a guy one time who tried to give me a wad of bills after the tour," Calder recalled. "He goes, 'Let me do something for your family.' I suggested he donate to the observatory instead, and he said he was going to, but he still wanted to give me some money too." (Calder declined the gift.)

He escorted me down through the control room, where the telescope is manipulated via computers and electronics, and inside the dome. It is a sort of cross between a science cathedral and a science factory, given over to one giant piece of industrial equipment aimed at the heavens: the HET. While it calls to mind some sort of futuristic vehicle poised to be launched into space, the telescope actually rotates on a fixed base, letting space come to it. Ninety-one hexagonal pieces form the primary mirror, a silver dish inside a steel cage; the light from the night sky hits this and then is relayed to an instrument above the mirror called the tracker. Itself a complex piece of machinery that took two years to build, the tracker can follow the image of a particular object as that image crosses the mirror. (Due to the rotation of the earth, the sky and its contents shift, relative to the position of the telescope, over the course of the night.)

Calder flipped a switch, engaging the air bearings below the telescope, and like a big hovercraft, it rose a few inches and then began to rotate. Moments later, a recording of squawking birds began to play, which is supposed to ward off real birds that might fly in through the ventilation shafts or through the top of the dome. Speaking over the sounds of the machinery and the squawking, Calder summarized the changes planned for the HET. "There's one camera up there right now," he said. "That whole top end is going to be replaced with a new top end that supports on the order of one hundred and fifty cameras, one hundred and fifty spectrographs." Those spectrographs, he told me, would be deployed, over the course of 140 nights, to record one million galaxies.

LATER THAT EVENING I CROUCHED on a platform below the Smith telescope, where Hill, MacQueen, and graduate student Josh Adams were bustling about a prototype of the instrument that is slated to be put on the HET. The instrument is called VIRUS, for Visible Integral-field Replicable Unit Spectrograph, and it will incorporate more than 145 copies of a simple spectrograph, each of which registers data from at least 246 fibers arrayed within something called an integral-field unit. Scattered on the floor of the platform were scissors and cables and screws and Allen wrenches, as if the scientists were fixing a motorcycle rather than testing a highly engineered piece of equipment, which was contained within an irregularly shaped black box about four feet wide and attached to the underside of the telescope's shaft by a metal harness. I watched as they checked the connections between the telescope and the box, cooled a crucial part with liquid nitrogen, and installed a glass plate, all in preparation for an observing run.

"There are four big links in the chain of observational astronomy," MacQueen explained as he connected the liquid nitrogen tank to the device. "Collecting the light, processing the light, detecting the light, and all the software side of analyzing the light." HETDEX will, if all goes smoothly, increase by about thirty-fold the area of sky that can be observed at any one time, speeding up the first three steps and thereby transforming the HET into a far more efficient and comprehensive cartographer of the skies. Rather than pointing at and shooting known objects, the upgrade will allow the research team to methodically obtain spectra from a giant celestial swath, taking information from more than 40,000 pixel-like sections of sky per exposure. Those will be captured by an 800-megapixel camera whose technology is similar to that of any vacationer's digital camera. "This is just a super-duper digital camera of sorts," MacQueen said.

Yet this is a camera that takes pictures of the remote past. Imagine a man in California trying to construct a map of the United States, using only the information brought to him by slow-moving (but long-lived) ants, creeping along at a hundred yards per year. The ants from New Jersey would have departed more than 50,000 years ago, before humans arrived in North America; the Midwestern ants some tens of thousands of years ago; the central California ants just a few years ago—so the map would record the East Coast as wild and uninhabited while the West Coast would contain highways and burger joints. This is the sort of map astronomers construct of the universe: The light that arrives from greater distances relays information about earlier eras.

If the big-bang model of the universe is correct, though, we cannot receive any messages from the earliest time periods, any more than our Californian mapmaker could learn about France. For some 380,000 years after the big bang, the universe was so hot that particles couldn't form atoms; electrons whizzed around every which way, and light couldn't travel any distance without colliding with those electrons. While this early fireball of a universe remains opaque to us, it is theorized that there must have been random variations in how densely its particles were distributed. Some areas were more dense than average, creating hot spots that drove particles out to cooler regions, which in turn caused waves to travel at the speed of sound. The result was a series of ripples through the sea of particles.

When the universe cooled down enough for atoms to form and light to escape, the oscillations ceased. At that particular moment in the history of the cosmos, the ripples were frozen into the density distribution of matter—so that now when astronomers map the heavens, they find the imprints of the early ripples in the way that galaxies are clustered. (It's as if you drew a pattern on a balloon and then inflated it. As the balloon expanded, the pattern would persist though its scale would change.) These imprints are what the HETDEX team plans to measure, to learn how they might have grown over time. By comparing the traces of the oscillations at different epochs, both with one another and with a pattern already discovered in something called the cosmic microwave background—which gives us a picture of the universe just after the light first escaped—they will be able to evaluate the changes in the scale of the universe and from that infer whether dark energy has made a constant contribution to the universe's expansion or has varied over the eons.

Some cosmologists have questioned whether a concerted push to do dark-energy experiments would be detrimental to the wider field of astronomy, hindering discoveries in other areas and relegating too many scientists to supporting roles on large-scale endeavors. After all, maybe there isn't any such thing as dark energy. Maybe no one really understands how gravity works. Or maybe the universe is not as homogenous as theorists have assumed. Maybe "dark energy" is the wrong way to frame the problem.

But even the possibility that dark energy might be a mistaken notion is no cause for pessimism, according to Gebhardt. "That would be the most exciting answer—if we're just so completely out to lunch," he said. "It means there is something fundamental about the nature of our universe that we don't understand. So whatever the answer is, it's going to be a fundamental shift in our understanding. In other words, it's not going to be a boring answer, and who knows where it will lead?"

THERESA BROWN

Perhaps Death Is Proud; More Reason to Savor Life

FROM *THE NEW YORK TIMES*

Theresa Brown, who came to nursing later in her life, has seen many patients die—and it is never easy.

AT MY JOB, PEOPLE DIE.

That's hardly our intention, but they die nonetheless.

Usually it's at the end of a long struggle—we have done everything modern medicine can do and then some, but we can't save them. Some part of their body, usually their lungs or their heart or their liver, has become too frail to function. These are the "good deaths," the ones where the family is present and knows what to expect. Like all deaths, these deaths are difficult, but they are controlled, unsurprising, anticipated.

And then there are the other deaths: quick and rare, where life leaves a body in minutes. In my hospital these deaths are "Condition A's." The "A" stands for arrest, as in cardiac arrest, as in this patient's heart has all of a sudden stopped beating and we need to try to restart it.

I am a new nurse, and recently I had my first Condition A. My patient, a particularly nice older woman with lung cancer, had been, as we say, "fine," with no complaints but a low-grade fever she'd had off and on for a couple of days. She had come in because she was coughing up blood, a problem we had resolved, and she was set for discharge that afternoon.

After a routine assessment in the morning, I left her in the care of a nursing student and moved on to other patients, thinking I was going to have a relatively calm day. About half an hour later an aide called me: "Theresa, they need you in 1022."

I stopped what I was doing and walked over to her room. The nurse leaving the room said, "She's spitting up blood," and went to the nurses' station to call her doctor.

Inside the room I found my patient with blood spilling uncontrollably from her mouth and nose. I remembered to put on gloves, and the aide handed me a face shield. I moved closer; I put my hand on her shoulder. "Are you in any pain?" I asked, as I recall, thinking that an intestinal bleed would be more fixable than whatever this was. She shook her head no.

I looked in her eyes and saw . . . what? Panic? Fear? The abandonment of hope? Or sheer desperation? Her own blood was gurgling in her throat and I yelled to the student for a suction tool to clear it out.

The patient tried to stand up so the blood would flow into a nearby trash can, and I told her, "No, don't stand up." She sat back down, started shaking and then collapsed backward on the bed.

"Is it condition time?" asked the other nurse.

"Call the code!" I yelled. "Call the code!"

The next few moments I can only describe as surreal. I felt for a pulse and there wasn't one. I started doing CPR. On the overhead loudspeaker, a voice called out, "Condition A."

The other nurses from my floor came in with the crash cart, and I got the board. Doing CPR on a soft surface, like a bed, doesn't accomplish much; you need a hard surface to really compress the patient's chest, so every crash cart has a two-by-three-foot slab of hard fiberboard for just this purpose. I told one of the doctors to help pick her up so I could put the board under her: she was now dead weight, and heavy.

I kept doing CPR until the condition team arrived, which seemed to happen faster than I could have imagined: the intensivists—the doctors who specialize in intensive care—the I.C.U. nurses, the respiratory therapists and I'm not sure who else, maybe a pulmonologist, maybe a doctor from anesthesia.

Respiratory took over the CPR and I stood back against the wall, bloody and disbelieving. My co-workers did all the grunt work for the condition: put extra channels on her IV pump, recorded what was happening, and every now and again called out, "Patient is in asystole again," meaning she had no heartbeat.

They worked on her for half an hour. They tried to put a tube down her throat to get her some oxygen, but there was so much blood they couldn't see. Eventually they "trached" her, put a breathing hole through her neck right into her trachea, but that filled up with blood as well.

They gave her fluids and squeezed bags of epinephrine into her veins to try to get her heart to start moving. They may even have given her adenosine, a dangerous and terrifying drug that can reverse abnormal heart rhythms after briefly stopping the patient's heart.

The sad truth about a true cardiac arrest is that drugs cannot help because there is no cardiac rhythm for them to stimulate. The doctors tried anyway. They went through so many drugs that the crash cart was emptied out and runners came and went from pharmacy bringing extras.

When George Clooney and Juliana Margulies went through these routines on *E.R.*, it seemed exciting and glamorous. In real life the experience is profoundly sad. In the lay vernacular of Hollywood, asystole is known as "flatlining." But my patient never had the easy narrative of the normal heartbeat that suddenly turns straight and horizontal. Her heartbeat line was wobbly and unformed, occasionally spiked in a brief run of unsynchronized beats, and at times looked regular, because chest compressions from CPR can create what looks like a real cardiac rhythm even though the patient is dead.

And my patient was dead. She had been dead when she fell back on the bed and she stayed dead through all the effort to save her, while blood and tissue bubbled out of her and the suction clogged with particles spilling from her lungs. Everyone did what she knew how to do to save her. She could not be saved.

The reigning theory was that part of her tumor had broken off and either ruptured her pulmonary artery or created a huge blockage in her heart. Apparently this can happen without warning in lung cancer patients. Only an autopsy could tell for sure, and in terms of the role I played in all this, it doesn't matter. I did the only thing I could do—all of us did—and you can't say much more than that.

I AM 43. I CAME to nursing circuitously, following a brief career as an English professor. Often at work in the hospital I hear John Donne in my head:

Death be not proud, though some have called thee
Mighty and dreadful, for thou art not so.

But after my Condition A I find his words empty. My patient died looking like one of the flesh-eating zombies from *28 Weeks Later*, and indeed in real life, even in the world of the hospital, a death like this is unsettling.

What can one do? Go home, love your children, try not to bicker, eat well, walk in the rain, feel the sun on your face and laugh loud and often, as much as possible, and especially at yourself. Because the only antidote to death is not poetry, or drama, or miracle drugs, or a roomful of technical expertise and good intentions. The antidote to death is life.

Evolutionists Flock to Darwin-Shaped Wall Stain

FROM *THE ONION*

With Darwin's theory of natural selection still duking it out in some areas of the United States against the forces of creationism and intelligent design, who else but the satiric newspaper The Onion *to put this great debate into perspective?*

DAYTON, TN—A STEADY STREAM OF DEVOTED EVOlutionists continued to gather in this small Tennessee town today to witness what many believe is an image of Charles Darwin—author of *The Origin of Species* and founder of the modern evolutionary movement—made manifest on a concrete wall in downtown Dayton.

"I brought my baby to touch the wall, so that the power of Darwin can purify her genetic makeup of undesirable inherited traits," said Darlene Freiberg, one among a growing crowd assembled here to see the mysterious stain, which appeared last Monday on one side of the Rhea County Courthouse. The building was also the location of the famed "Scopes Monkey Trial" and is widely considered one of Darwinism's holiest sites. "Forgive me, O Charles, for ever doubting your Divine Evolution. After seeing this miracle of limestone pigmentation with my own eyes, my faith in empirical reasoning will never again be tested."

Added Freiberg, "Behold the power and glory of the scientific method!"

Since witnesses first reported the unexplained marking—which appears to resemble a 19th-century male figure with a high forehead and large beard—this normally quiet town has become a hotbed of biological zealotry. Thousands of pilgrims from as far away as Berkeley's paleoanthropology department have flocked to the site to lay wreaths of flowers, light devotional candles, read aloud from Darwin's works, and otherwise pay homage to the mysterious blue-green stain.

Capitalizing on the influx of empirical believers, street vendors have sprung up across Dayton, selling evolutionary relics and artwork to the thousands of pilgrims waiting to catch a glimpse of the image. Available for sale are everything from small wooden shards alleged to be fragments of the "One True *Beagle*"—the research vessel on which Darwin made his legendary voyage to the Galapagos Islands—to lecture notes purportedly touched by English evolutionist Alfred Russel Wallace.

"I have never felt closer to Darwin's ideas," said zoologist Fred Granger, who waited in line for 16 hours to view the stain. "May his name be praised and his theories on natural selection echo in all the halls of naturalistic observation forever."

Despite the enthusiasm the so-called "Darwin Smudge" has generated among the evolutionary faithful, disagreement remains as to its origin. Some believe the image is actually closer to the visage of

Stephen Jay Gould, longtime columnist for *Natural History* magazine and originator of the theory of punctuated equilibrium, and is therefore proof of rapid cladogenesis. A smaller minority contend it is the face of Carl Sagan, and should be viewed as a warning to those nonbelievers who have not yet seen his hit PBS series *Cosmos: A Personal Voyage.*

Still others have attempted to discredit the miracle entirely, claiming that there are several alternate explanations for the appearance of the unexplained discoloration.

"It's a stain on a wall, and nothing more," said the Rev. Clement McCoy, a professor at Oral Roberts University and prominent opponent of evolutionary theory. "Anything else is the delusional fantasy of a fanatical evolutionist mindset that sees only what it wishes to see in the hopes of validating a baseless, illogical belief system. I only hope these heretics see the error of their ways before our Most Powerful God smites them all in His vengeance."

But those who have made the long journey to Dayton remain steadfast in their belief that natural selection—a process by which certain genes are favored over others less conducive to survival—is the one and only creator of life as we know it. This stain, they claim, is the proof they have been waiting for.

"To those who would deny that genetic drift is responsible for a branching evolutionary tree of increasing biodiversity amid changing ecosystems, we say, 'Look upon the face of Darwin!'" said Jeanette Cosgrove, who, along with members of her microbiology class, has maintained a candlelight vigil at the site for the past 72 hours.

"Over millions of successive generations, a specific subvariant of one species of slime mold adapted to this particular concrete wall, in order to one day form this stain, and thus make manifest this vision of Darwin's glorious countenance," Cosgrove said, overcome with emotion.

"It's a miracle," she added.

About the Contributors

THERESA BROWN, RN, is a floor nurse in a medical oncology unit. She is writing a book for HarperStudio titled *Critical Care: A Nurse's First Year*, an autobiographical account of her first year of nursing. Ms. Brown has written for the "Cases" column in the *New York Times*, where this essay first appeared. She is also a regular contributor to the *New York Times* blog "Well." In addition to her nursing degree, Ms. Brown holds a Ph.D. in English from the University of Chicago and taught writing for three years at Tufts University.

"I wrote this essay after the event it describes because I hoped that writing would help contain the experience," she explains. "'Look,' I could say to myself, 'this story takes up three pages, 1,300 words, not an infinitude of fear and sadness.' Once I finished writing the piece, I thought, 'This is good,' and sent it to the *New York Times*. Writing about my patient's sudden and gruesome ending made her death easier to bear, but I'm still haunted by the experience, especially whenever I'm involved in a code."

MARINA CORDS is a professor of ecology, evolution, and environmental biology at Columbia University in New York. Her research focuses on the social behavior, ecology, and reproduction of nonhuman primates, especially Old World monkeys. She has led a field study of guenon monkeys in the Kakamega Forest in western Kenya since 1979, in which she combines behavioral observations with laboratory-based investigations of DNA and hormones. She also works with local people who formed a community-based organization to foster forest conservation through education.

"Primates don't have a monopoly on social cooperation," she says, "but what makes them especially interesting to me is that their social lives are long. Natural selection acts on lifetime reproductive success, and a lot can happen in a primate's lifetime, not only to that individual, but to its whole group. This is why I've traipsed off to the forest for decades now: to understand the general patterns, one has to witness a lot of lifetimes. I get caught up in the moment of being a voyeur in another social world, but the real satisfaction comes in trying to relate the patterns to interesting theory."

GREGG EASTERBROOK is the author of numerous books including *The Progress Paradox* (Random House, 2004) and *Sonic Boom* (Random House, 2009). He is a contributing editor to *The Atlantic*, to *The New Republic*, and to *The Washington Monthly*. His Web site is www.greggeasterbrook.com.

"In March 2009, an asteroid estimated at 30 meters wide, designated 2009 DD45, passed between the Earth and the moon," he reports. "Although the asteroid was never on a collision course with Earth, it came disturbingly close—missing by 48,800 miles, not much above the orbit of communications satellites. Had the space object struck ground, it could have obliterated a city the size of Paris. The asteroid arrived without any warning—indeed, was not spotted until hours before it passed the planet. Had it been on a collision course, Earth would have been helpless."

MARTIN ENSERINK, a native of the Netherlands, became a science journalist after obtaining a master's degree in biology in 1989. In 1999, he joined the news staff of *Science* magazine at its headquarters in Washington, D.C. He became a contributing correspondent for *Science* in 2004, and currently lives in Amsterdam and Paris. Enserink specializes in infectious diseases and research policy. He has covered many disease outbreaks and has written extensively about neglected diseases, biodefense, and pandemic threats.

"For this story, I wanted to find out what happened to golden rice," he states, "which seemed to have disappeared from public view after making global headlines a decade ago. What I discovered was a fascinating personal and scientific saga—and a prime example of the often intense interplay of politics and research.

"In the end, the story isn't just about transgenic rice. Perhaps the most poignant conclusion is that many scientists have given up on the use of genetic modification as a way to help the world's poorest—not for scientific, but for political, reasons."

ATUL GAWANDE is a staff writer for *The New Yorker* and general and endocrine surgeon at Brigham and Women's Hospital. He is also an associate professor at Harvard Medical School and the Harvard School of Public Health. In 2006, Dr. Gawande received a MacArthur Award for his research and writing. He is the author of *Complications: A Surgeon's Notes on an Imperfect Science*, which was a 2002 finalist for the National Book Award, and *Better: A Surgeon's Notes on Performance*.

"Some readers, upon finishing 'The Itch,' have naturally wondered what has happened with M.'s itch," he remarks. "As of early 2009, it's still no better, though she and her neurologist haven't yet tried the reverse-mirror treatment idea. I've heard from other patients with similar symptoms who've experimented with the approach and found gratifying success, and I've passed along their experiences. But the treatment remains purely speculative and awaits a proper trial. There is also some ambiguity in my role as a practicing doctor who does journalism. The minute I am actually devising new treatments for people who aren't my patients, especially as a general surgeon and not a neurologist, it feels like I'm crossing a line. So I have been happy to bring M. and others the idea. But I have also been careful not to treat them as if I were their doctor."

JOHN HORGAN is a freelance journalist who teaches at and directs the Center for Science Writings at Stevens Institute of Technology, Hoboken, New Jersey. A former senior writer at *Scientific American*, he has also written for the *New York Times*, *National Geographic*, *Discover*, *Time*, and other publications. His books include *The End of Science* (1996), *The Undiscovered Mind* (1999), and *Rational Mysticism* (2003).

"During the U.S. invasion of Iraq in March 2003," he writes, "Frank Geer, an Episcopal priest in Garrison, New York, my hometown, asked me to speak in his church. In my talk, titled 'Is War in Our Genes?,' I reviewed findings about the roots of warfare, which extend deep into human prehistory. I concluded on an upbeat note: 'We will abolish war someday. The only question is how, and how soon.' When I asked if anyone agreed with me, only a dozen of the 60 people present raised their hands. My *Discover* article reflects my ongoing attempt to persuade others, and myself, that we're smart and decent enough to solve the problem of war."

JENNIFER KAHN has been a contributing editor at *Wired* magazine since 2003, and a feature writer for *The New Yorker, National Geographic, Outside, Discover, Mother Jones*, and the *New York Times*, among others. She is currently a visiting fellow at the UC Berkeley Graduate School of Journalism.

"'A Cloud of Smoke' grew out of an article I read about the death of James Zadroga, a 34-year-old policeman who'd joined the rescue effort at Ground Zero and died four years later with terrible lung damage," she explains. "At the time, I didn't realize that Zadroga was famous as the first 'dust death' connected to 9/11. What puzzled me was the medical debate: one pathologist said the damage came from inhaling toxic dust at the site; another, the Chief Medical Examiner of New York, said it was caused by drug use—from shooting up pills that had been crushed in solution. It wasn't immediately clear who was right, and that alone seemed strange. How could drug abuse and inhaled dust even look alike, postmortem?"

Alex Kotlowitz is the author of three books, including *There Are No Children Here*, which the New York Public Library selected as one of the 150 most important books of the twentieth century. A regular contributor to *The New York Times Magazine* and public radio's *This American Life*, he teaches writing at Northwestern University. His journalism honors include the George Foster Peabody Award, the Robert F. Kennedy Journalism Award, and the George Polk Award. His play *An Unobstructed View* (coauthored with Amy Dorn) premiered in Chicago in June of 2005.

"I came to write about Ceasefire out of my own personal experiences," he recalls. "The one constant in the lives of the kids I wrote about in *There Are No Children Here* was the violence, and, sadly, two boys I knew well were shot and killed in separate incidents. A public health model seems like the most sensible (and creative) way to think about this epidemic, and as Gary Slutkin argues, the most sensible way to attack it. In my twenty years writing on urban poverty and the accompanying violence, Slutkin's approach has felt the freshest and the most promising."

Jennifer Margulis, Ph.D., worked in nonprofit development in Niger, West Africa, from 1992 to 1993, which is when she first saw the last herd of West African giraffes. She went back to Niger with her family as a Fulbright Fellow in 2006–2007 and taught nineteenth-century American literature to students at the University of Abdou Moumouni, and again on assignment for *Smithsonian* in 2008. She is the daughter of evolutionary microbiologist Lynn Margulis. She has also written about large land mammals for *Wildlife Conservation* magazine and the *Christian Science Monitor.* Learn more about her at www.jennifermargulis.net.

"When you tell people you're going to Niger," she says, "most say, 'Oh, you mean Nigeria?!' Recently a travel editor told me he liked my piece about Nigeria's giraffes. But Niger is very different from its richer cousin to the south. One of the world's poorest countries, Niger is a dusty, landlocked place where temperatures climb to 130°F in the shade. Walking with the last West African giraffes on the baking sand in the desolate land where they survive filled me with wonder. I encourage anyone with a hankering for travel or an interest in land mammals to visit Niger's giraffes."

JINA MOORE is a freelance journalist who covers international affairs, human rights, and science from the United States and Africa. Her general-interest features appear regularly in the *Christian Science Monitor.* Previously, she was an editorial assistant at *Harper's* and helped run the Nieman Conference on Narrative Journalism at Harvard University. You can read her work and follow her blog at www.jinamoore.com.

"I started this project in 2006," she says, "when the national conversation about torture was obsessed with ticking time bombs, legalese, and hair-splitting rationalizations. I remembered an essay by Jean Améry, a Frenchman who resisted the Nazis, that influenced me as a teenager. Unsentimental and spare, he tried to explain how the body and mind react to torture. I wrote this piece to see if thinking about torture physiologically, as Améry did, influences how we think about it morally. I also wanted to confront torture's inherent brutality, an element we seemed too willing to gloss over in our collective conversation."

J. MADELEINE NASH is a freelance journalist based in San Francisco. Her articles have appeared in *Time*, *Smithsonian*, the *New York Times*, and *High Country News.* She is the author of a book, *El Niño: Unlocking the Secrets of the Master Weather-Maker*, and a three-time winner of the American Association for the Advancement of Science magazine writing award.

"Perhaps it's because I majored in history at college," she explains, "but I've long been drawn to scientific explorations of the past. In fact, you can put the term 'paleo' in front of almost anything, and it's liable to grab my attention. So in 2007, when I heard scientists talk about the sudden warming that occurred 55.5 million years ago—the so-called Paleocene-Eocene Thermal Maximum—I knew it was something I wanted to write about. A couple of months later, Scott Wing and I started talking about his work in the Big Horn Basin. 'You really *have* to come,' he said."

A native of Washington, D.C., KAREN OLSSON lives in Austin, Texas. She is a senior editor for *Texas Monthly* and the author of the novel *Waterloo.*

"One of the perks of reporting in Texas," she relates, "is stumbling across unexpected cultural collisions, i.e., a couple of astronomers driving from one west Texas town to the next to promote a dark energy experiment to locals. The HETDEX scientists' fund-raising efforts served as a reminder of how much extrascientific work is required to bring such a big, expensive experiment to fruition. Presumably it's not the sort of thing they studied in graduate school, yet public relations seemed to come fairly naturally to Gary, Karl, and their collaborators, thanks to their enthusiasm for the project.

"The McDonald Observatory itself is a beautiful spot; the combination of natural beauty and helpful interview subjects made this article a really enjoyable one to research. Since it appeared, the HETDEX team has passed a couple of reviews and obtained more than two-thirds of the needed funds. They plan to start making observations in 2011."

DENNIS OVERBYE is a science correspondent for the *New York Times*. He is the author of two books, *Lonely Hearts of the Cosmos: The Scientific Search for the Secret of the Universe* (Little, Brown, 1999) and *Einstein in Love: A Scientific Romance* (Penguin, 2001). He lives in New York City with his wife, the writer Nancy Wartik, and his daughter, Mira.

He writes: "There is an old comic strip—which one I don't remember—in which the characters are sitting looking at the night sky and musing on its meaning, when one of them suddenly says something like 'Oops, I think I just sprained my imagination.' This is an article about cosmologists who have sprained their imaginations.

"In recent years, astronomers, often against their will, have been forced to confront the notion of multiple universes in order to make sense of how various discoveries and theories fit together. Among them are string theory, which seems to have an inexhaustible supply of solutions, and the discovery that the expansion of the universe is speeding up, which if it continues could provide the stage for an endless recurrence of big bangs and universes.

"How these alternate universes relate to each other is not clear. Why should we care what goes on in them? Are they parallel, like a stream that keeps splitting and re-splitting? Do they recur sequentially from now until eternity or are they different zones—sometimes called pocket universes—with a multiverse that already represents eternity? Can you get there from here if 'there,' say, has nine dimensions and 'here' has three or four?

"All these questions have served to underscore some fundamental issues about our own universe that have remained murky, like just how it started and why time only seems to run in one direction. A few years ago I started to hear people mentioning 'Boltzmann brains' as a problem in these eternal vistas. The basic idea is that if you are going to have universes spontaneously occur from time to time or place to place, it is much easier and cheaper, from a probability point of view, to have pieces of a universe or just naked brains pop into existence than a whole universe.

"So why aren't we an illusion? How do we compare the likelihoods of events in different kinds of universes? How do we measure eternity? I just sprained my imagination."

ANNIE MURPHY PAUL is a book author and magazine journalist who contributes to *The New York Times Magazine*, *The New York Times Book Review*, and *Slate*, among other publications. She is the author of *The Cult of Personality: How Personality Tests Are Leading Us to Miseducate Our Children, Mismanage Our Companies, and Misunderstand Ourselves*, published in 2004 by the Free Press. A former senior editor of *Psychology Today* magazine, she is the recipient of a Rosalynn Carter Mental Health Journalism Fellowship. She is a graduate of Yale College and the Columbia University Graduate School of Journalism.

She writes: "I first encountered the question of whether fetuses can feel pain while researching my current book project, about changing scientific and social notions of pregnancy. I was fascinated by the scientific debate over the mechanics of pain perception, but even more interesting to me were the philosophical questions: What is the relationship between humanness and the ability to feel pain? What responsibility do we have to prevent pain, wherever it might occur? We know so much more than we once did about the fetus, yet the fetus remains mysterious, a blank screen on which we project our beliefs and assumptions."

CATHERINE PRICE is a contributing editor at *Popular Science* whose freelance work has appeared in publications including the *New York Times*, *Salon*, *Slate*, *The Washington Post Magazine*, and *Mother Jones*, among others. A graduate of Yale and UC Berkeley's Graduate School of Journalism, she is the author of an upcoming travel book, *101 Places Not to See Before You Die*, and a recipient of the 2008 Middlebury Fellowship in Environmental Reporting. She particularly enjoys writing about experiences that are painful in the moment but amusing afterwards, including bicycling across the United States, traveling halfway across the Pacific by sailboat, driving a formula 1 race car, and teaching hip-hop aerobics in Beijing. You can see more of her work at catherine-price.com.

"Writing this article made me paranoid about sharing personal information with retailers—you just don't know what it'll be used for," she explains. "But I realized my paranoia had gone too far when I refused to give my phone number to an employee at Michael's Crafts after I'd dropped off some pictures to be framed. 'What do you need it for?' I asked suspiciously, imagining they were going to sell my number to a data aggregator that kept track of people's craft shop purchases. 'Um, I was going to use it to call you when your order was ready for pickup,' he said. 'But I guess I could just send you a letter instead. You know, with, like, a wax seal?'

"I gave him my phone number."

David Quammen is the author of eleven books, including *The Song of the Dodo* and *The Reluctant Mr. Darwin*, and general editor of a new, illustrated edition of Charles Darwin's *On the Origin of Species.* Quammen is a contributing writer for *National Geographic* and writes also for *Harper's* and other magazines. He lives in Montana with his wife, Betsy Gaines, a conservationist.

"Since this article was written," he notes, "further evidence has confirmed the extraordinary fact that Devil Tumor is a genuinely transmissible line of cancer cells, not a sequence of cancers caused by a transmissible virus. Hannah V. Siddle, of the University of Sydney, and her colleagues have published that evidence in *Proceedings of the National Academy of Sciences* (vol. 104, no. 41). Meanwhile the devil itself continues to suffer not just from its cancer epidemic but also from ongoing habitat losses, due to timber harvesting and other practices, on the island of Tasmania."

Elizabeth Royte is a Brooklyn-based freelance writer. A frequent contributor to *The New York Times Book Review* and a contributing editor for *OnEarth* magazine, she has written about the environment for *The New York Times Magazine, Outside, National Geographic, Mother Jones,* and other national publications. Her most recent books are *Bottlemania: How Water Went On Sale and Why We Bought It* and *Garbage Land: On the Secret Trail of Trash.*

She writes: "My interest in recycling wastewater into drinking water grew from research I conducted for my most recent book, *Bottlemania,* which takes a hard look at tap water as well as bottled. While I was at first blush skeptical about drinking treated sewage—who wouldn't be?—I came to see the value of treating wastewater as a resource, especially in chronically dry areas. Since this article appeared, several other U.S. municipalities have started to investigate similar water-reuse schemes."

Oliver Sacks is a professor of neurology and psychiatry at Columbia University and was recently appointed Columbia's first University Artist. He is a frequent contributor to *The New Yorker* and *The New York Review of Books*, and is best known for his neurological case studies, published in books such as *The Man Who Mistook His Wife for a Hat* and *Musicophilia*. His book *Awakenings* was the basis for the feature film starring Robin Williams and Robert De Niro. For more information, see www.oliversacks.com.

"I first read *Journey Round My Skull* when I was 13 or 14," he recalls. "My parents, who had both trained in neurology as young doctors, were fascinated by it, and it was their copy I read. The book had been translated into many languages after its initial 1938 publication in Hungary, and it attracted considerable attention. The brain surgery scene in the 1946 film *A Matter of Life and Death* was inspired by Karinthy's description of his own surgery. But by the 1960s, the book was out of print.

"In 1999, when *The New York Review of Books* began publishing their wonderful series of forgotten classics, I immediately suggested that they republish *Journey Round My Skull*. So in 2008, I was delighted when they approached me to write this introduction, which appeared first as an essay in *The New York Review of Books*."

John Seabrook is a staff writer at *The New Yorker*. He is the author of three books, *Deeper*, *Nobrow*, and the recently published collection, *Flash of Genius*, the title story of which was the basis for the film of the same name starring Greg Kinnear.

"I've been interested in talking computers ever since I saw *2001: A Space Odyssey*, when I was eight," he explains. "In approaching this subject as a writer, I was interested in the way in which movie fantasies about technology can both inspire development and create false expectations that may retard development. I think both things happened with speech recognition."

Margaret Talbot has been a staff writer for *The New Yorker* since 2003. She lives in Washington, D.C., with her husband, writer Arthur Allen, and their children, Isaac and Lucy, both of whom can now do excellent imitations of Alex the parrot.

"One of the ironies that strikes me when I look back on this story," she reflects, "is that researching sociability and communication in animals can be a fairly lonely pursuit. Irene Pepperberg's work with Alex was something that earned her a lot of skepticism and isolated her from other scientists to some degree, and getting parrots to talk in a meaningful way required her to devote a huge amount of time and emotional energy to cultivating a relationship with them. And African grays are needy, sensitive birds—the Woody Allens of the avian world."

SALLIE TISDALE is the author of *Women of the Way* and several other books. Her essays have appeared in many different journals and magazines, including *Harper's*, *Antioch Review*, and *Tricycle*.

"Since 'Twitchy' was published," she reveals, "I experienced an altogether new dental adventure when a new gold crown and a new mercury filling became galvanically active. The ensuing electric shocks created the weirdest and most alarming dental pain I've known. Dr. B. cured me. I love my dentists very much."

GARY WOLF is a contributing editor at *Wired* magazine and the author of *Wired—A Romance*, and *Dumb Money* (with Joey Anuff).

"I thought of Piotr Wozniak as a hero before I met him," he says. "A few years ago, while learning Spanish, I stumbled across Wozniak's confident and rather eccentric essays. I tried his method, and my pace of learning dramatically improved. Who was this person? And why weren't more people using his technique? These questions took me on an intellectual adventure back to the founding era of experimental psychology, and a physical adventure to a tiny Polish town on the edge of the Baltic, where Wozniak lives a life entirely devoted to personal optimization, with himself as the chief experimental subject."

DAVID WOLMAN is a contributing editor at *Wired* and the author of two works of nonfiction: *A Left-Hand Turn Around the World* and *Righting the Mother Tongue*. He has written for *Newsweek*, *Discover*, *New Scientist*, *Outside* and numerous other publications, and was a Fulbright journalism fellow in Japan. He now lives in Portland, Oregon.

Permissions

"The Final Frontier" by Karen Olsson. Copyright © 2008 by *The Texas Monthly.* Reprinted by permission.

"Evolutionists Flock to Darwin-Shaped Wall Stain" by the *Onion*. Reprinted with permission of *THE ONION.* Copyright © 2008 by ONION, INC. www.theonion.com.

"Big Brain Theory: Have Cosmologists Lost Theirs?" by Dennis Overbye. From the *New York Times*, January 15, 2008. Copyright © 2008 by the *New York Times.* All rights reserved. Used by permission and protected by the Copyright Laws of the United States. The printing, copying, redistribution, or retransmission of the Material without express written permission is prohibited.

"The First Ache" by Annie Murphy Paul. Copyright © 2008 by Annie Murphy Paul. All rights reserved. Reprinted by kind permission of the author. First published in *The New York Times Magazine.*

"The Anonymity Experiment" by Catherine Price. Copyright © 2008 by *Popular Science.* Reprinted by permission.

"Contagious Cancer" by David Quammen. Copyright © 2008 by *Harper's Magazine.* All rights reserved. Reproduced from the April issue by special permission.

"A Tall, Cool Drink of . . . Sewage?" by Elizabeth Royte. Copyright © 2008 by Elizabeth Royte. All rights reserved. Reprinted by kind permission of the author. First published in *The New York Times Magazine.*

"A Journey Inside the Brain" by Oliver Sacks. © 2008 by Oliver Sacks. First published in *The New York Review of Books.* Reprinted with the permission of The Wylie Agency, LLC.

"Hello, HAL" by John Seabrook. Copyright © 2008 by John Seabrook. All rights reserved. Reprinted by kind permission of the author. First published in *The New Yorker.*

"Birdbrain" by Margaret Talbot. © 2008 by Margaret Talbot. First published in *The New Yorker.* Reprinted with the permission of The Wylie Agency, LLC.

"Twitchy" by Sallie Tisdale. Copyright © 2008 by the *Antioch Review*, Inc. First appeared in the *Antioch Review*, vol. 66, no. 2. Reprinted by permission of the Editors.

"Want to Remember Everything You'll Ever Learn? Surrender to This Algorithm" by Gary Wolf. Copyright © 2008 by Gary Wolf. All rights reserved. Reprinted by kind permission of the author. First published in *Wired.*

"The Truth About Autism" by David Wolman. Copyright © 2008 by David Wolman. All rights reserved. Reprinted by kind permission of the author. First published in *Wired.*

A Note from the Series Editor

Submissions for next year's volume can be sent to:

Jesse Cohen
c/o Editor
The Best American Science Writing 2010
HarperCollins Publishers
10 E. 53rd St.
New York, NY 10022

Please include a brief cover letter; manuscripts will not be returned. Submissions can be made electronically and sent to jesseicohen@netscape.net.